tic Ocean
Monte Cristi
n
e Laferrière
t-Raphael
ne
ille
DOMINICAN REPUBLIC
Santo Domingo
USA
Atlantic
Cuba
Area of detail
n Sea
Venezuela
N
W
E
S

INTERVENTION!

The Americans in Haiti, 1915-1934

Commander Richard L. Schreadley,
U. S. Navy (Retired)

INTERVENTION!

The Americans in Haiti, 1915-1934

Commander Richard L. Schreadley,
U. S. Navy (Retired)

EVENING POST BOOKS

Published by
Evening Post Books
Charleston, South Carolina

First edition

Editor: John M. Burbage
Designer: Gill Guerry

First printing 2017
Printed in the United States of America

A CIP catalog record for this book has been applied
for from the Library of Congress.

ISBN: 978-1-929647-31-6

DEDICATION

To the men and women of the armed forces, who serve their country on delicate and dangerous missions all over the world. They are too little recognized for what they accomplish, and too often shoulder blame when hapless "nation-building" goes astray, as it did in Haiti many years ago, in Vietnam in the 1960s and 1970s, and in the ongoing maelstrom that is the Middle East.

CONTENTS

FOREWORD

On July 28, 1915, U.S. Marines and blue jackets from the armored cruiser *Washington* were landed at Port-au-Prince, Haiti, for the purpose of "preventing further rioting and for the protection of foreigners' lives and property." Thus began a military occupation of nearly 20 years.

A brief history of Haiti, the events leading up to the U.S. intervention, the treaty that "legalized" it, the structure and operation of the treaty regime, and steps taken to end it are recounted in the pages that follow.

Descriptive passages are based on first-hand accounts wherever possible. Principal sources are archival materials from the U.S. Department of State and the Navy. Others include the Adm. William B. Caperton papers in the Library of Congress, an unpublished manuscript by Capt. Edward L. Beach Sr., and contemporary newspaper and magazine articles.

Impressive and instructive are extra-governmental individuals and groups that were involved in shaping U.S. policy there. Both Haitian and U.S. politicians along with missionaries, civil libertarians, businessmen, journalists, bankers and philosophers played significant roles in how the intervention progressed and ended.

Prior to the intervention, Haiti was like pre-World War II Czechoslovakia — a far away land of which we knew little. Afterwards, however, self-declared experts emerged everywhere offering criticism and unsolicited advice to those who actually bore responsibility for formulating and carrying out U.S. policy. Some criticism was warranted and some of the advice was sound, but much of what was said and written about Haiti during the intervention was not well founded on fact.

Haiti's occupation in the 1920s was a political issue *sans pareil* in part because of a U.S. swing toward isolationism and a between-the-world-wars distrust of the military due in part by the racial issues involved. America was not yet prepared to grant its own large black population an equal status, but neither was it tolerant of reported white injustice toward Haitian blacks.

Yellow journalism flourished, and isolated instances of wrongdoing by

Americans in Haiti were seized upon and presented as a general policy of repression and brutality toward blacks enforced by U.S. Marine bayonets and machineguns. Such charges echoed in the radical press of Latin America and Europe. Haiti was a convenient stick with which to batter the United States.

Each successive U.S. administration recognized the dissension at home and the damage to U.S. prestige abroad caused by the continued occupation of Haiti. Withdrawal, however, posed serious questions of conscience as well as concern that failure to effect needed reforms there might require a second intervention. U.S. officials were loath to recommend precipitate retreat in the face of explicit threats to Haitians who cooperated with the Occupation and to their property.

There appeared to be no honorable way out, and in light of Haitian history since the ending of the occupation, one is tempted to conclude that there was in fact no exit without a clear admission of failure.

It is difficult to resist comparing the U.S. intervention in Haiti in 1915 with the infinitely more costly and tragic U.S. intervention in Vietnam in the 1960s and 1970s. In neither country was there a clearly defined U.S. national interest at stake. In neither country were there significant U.S. investments to defend. The strategic considerations often cited for the interventions apparently did not exist when the involvements began.

The subject of Caribbean naval bases and coaling stations was a dead letter by 1915 as a result of changes wrought by the Spanish-American War. The "domino theory" in Southeast Asia and the strategy of containment were largely discredited by 1965 in view of the breakup of monolithic communism and new alignment of world powers.

Both in Haiti and Vietnam high moral principles were invoked to justify intervention: America must save Haiti from the chaos and anarchy brought on by constant revolutions, and America must ensure that the free people of Vietnam have the right to choose their own form of government. In both countries, America signally failed to nurture the democratic institutions it struggled to implant in those unhappy lands. Why? A frequently voiced complaint was that America simply bet on the wrong horses. U.S.-backed Haitian and Vietnamese governments were attacked for corruption, ineptness and unwillingness to embrace the "American way."

More often forgotten than remembered are President Woodrow Wilson's thoughts on the subject of U.S. interventions: "Only an honest and efficient government deserves support. The Government of the United States could not justify the expenditure of money and the sacrifice of American lives in support of any other kind of government."

It seems probable, however, that the root causes of U.S. failures in Haiti and Vietnam are more fundamental to the human condition than mere errors in political judgment. Freedom, progress and political evolution are seldom if ever successfully imposed from without.

PART ONE:

HAITI PRIOR TO THE INTERVENTION

What is past is prologue. Study the past.

— Pedestal inscription, National Archives, Washington D.C.

1

FROM COLUMBUS TO THE HAITIAN REVOLUTION

So lovable, so tractable, so peaceable are these people that I swear to your Majesties there is not in the world a better nation nor a better land.

— Christopher Columbus to Ferdinand and Isabella.

However it is approached — by sea, by air or overland on near impossible roads from the Dominican Republic — Haiti's outstanding geographic feature is mountains. The name "Haiti" itself is of native origin, probably meaning "the high place." Other than a few stone artifacts, the name remains a rare reminder of the Ciboney, Awawak and Carib people who once lived there.

Hammond's World Atlas credits Haiti with 10,714 square miles of territory, about 80 percent of which is mountainous. The Cordillera Central range — or Cibao Mountains — stretches across the northern peninsula with peaks at 8,000 feet above sea level. In Haiti's southeastern region, the La Selle, and in the southern peninsula, the La Hotte, form the backbones of the country's other mountain chains. Seen from afar they are not unlike some parts of the great Appalachian range in the eastern United States.

Haiti has three principal rivers and three lowlands. The largest river — the Artibonite — empties into the sea near Gonaives on the west coast after passing through a wedge-shaped valley of the same name. The second river — La Grande Anse — drains a part of the rugged southern peninsula. Its mouth is near Jeremie. The third — Trois Rivieres — flows into the sea on the north coast near Port de Paix.

Santo Domingo is said to be the sixth point of land charted by Columbus on his voyage of discovery of the New World in 1492.

Between the mountains and the sea in the north lies the Plaine du Nord, the most fertile region in the country. Between the central and southern ranges, extending from the capital — Port-au-Prince — into the Dominican Republic, is the great semi-arid plain known as the Cul de Sac.

The climate is tropical but Haiti, particularly in the north, is cooled by prevailing sea breezes.

Haitians are heir to only about a third of the island. Except for relatively brief periods in the 19th century when it was conqueror of the whole, Haiti has been the uneasy bedfellow of the Dominican Republic, occupier of the eastern two-thirds of the island of Santo Domingo, also called Hispaniola.

The island they share, however, is one of the most strategic locations in the Western Hemisphere. From the southeast it commands the entrance to the Gulf of Mexico. More important, the two great ocean highways linking North

America and Europe to the Panamanian isthmus — the Mona Passage on the east and the Windward Passage on the west — stream by its coast, a fact not unnoticed by naval strategists from Morgan to Mahan. The island's nearest neighbors in the Greater Antilles chain are Cuba to the northwest and Puerto Rico to the southeast. New York City is about 1,300 miles north.

Little is known of the original inhabitants. They left no monuments and no written records. Those who welcomed Columbus called themselves *Tainos* (the Good People), believed to be of Arawak stock. The story of how and when they and other tribes came to Haiti remains shrouded in the past.

Santo Domingo is said to be the sixth point of land charted by Columbus on his voyage of discovery of the New World, and his landfall probably was near Mole St. Nicholas on Dec. 5, 1492. From there his tiny squadron sailed eastward along the coast until disaster struck on Christmas Day near Cap Haitian. His flagship, *Santa Maria*, ran aground and efforts to re-float her failed.

From the ship's timbers a small fortification was built on the nearby shore. Salvaged goods provided a meager supply of stores, and with great difficulty the crew dragged ashore a few of the ship's cannons, powder and rounds. Thirty-eight crewmen volunteered to stay when Columbus set sail with his remaining ships for a return to Spain. The stockade and shelter built from the wreckage of the *Santa Maria* was christened "La Navidad," commemorating the day of the grounding. It is not known what transpired among the men during the admiral's absence. On his return the following year, he found no survivors.

Columbus returned to La Isla Espanola (later somewhat Latinized as Hispaniola) with settlers and great plans for making the island the hub for Spain's exploration and exploitation of the New World. Chance, and perhaps geographical imperative, led to the establishment of the principal city and emporium at Santo Domingo, on the southeastern coast, which was more convenient for Spanish fleets sailing to Central and South America.

The city was settled in 1496 by Columbus' brother Bartholomew and is considered the first permanent settlement in the Western Hemisphere. The shifting of Spanish interest to the eastern part of the island was critical in Haiti's later development.

Much to the disgust of the conquistadors, the few placer mines where the natives gleaned bits and pieces of gold were soon panned out. But sugar cane — introduced on Columbus' second New World voyage from the Canary Islands — thrived in the rich soil and climate of Hispaniola. Its intensive cultivation hastened the extinction of the island's aborigines, already decimated by European diseases and not well suited for heavy field labor.

Of a native population of an estimated 300,000 in 1492, only 60,000 remained in 1508, and 40 years later only 500 were left. Bishop Bartoleme de Las Casas was touched by the horrors inflicted on the natives and upon his urging the Crown consented to import African slaves to the island. The first of them arrived in 1512 and worked mostly in the eastern sector. For more than a century afterward the western part (Haiti) was virtually deserted.

The hegemony of Spain in Hispaniola, as in other parts of the New World, did not go unchallenged. In 1629 French and English survivors of a Spanish assault on their "illegal" settlement on St. Christopher in the Leeward Islands, escaped in open boats to the small island of Tortuga, off the northwestern coast of Hispaniola. Dutch refugees later joined them. From their custom of cooking meat over spits call *"boucans,"* they came to be known as *"boucaniers"* — a name that clung to them in slightly corrupted form when they turned their hunting instincts to the sea.

Continually harassed by the Spanish, the buccaneers organized a society called "Brethren of the Coast." With tacit support of English, Dutch and French warships and privateers, they proceeded to terrorize the Spanish Main. Their principal settlement on Tortuga became one of the most prosperous, dissolute and bawdy of colonies in the West Indies. A chronic shortage of women was from time to time redressed by the purchase of African women from traders or by abduction, along with a few white women from the Spanish settlements. From these unions came the Creole and mulatto people destined to play prominent roles in Haitian history.

From about the middle of the 17th century, French influence predominated on Tortuga and gradually spread along the adjacent coast of western Hispaniola. Small settlements and farms sprang up along the riverbanks and spread into interior valleys, foreshadowing the great French plantation society that flowered in the 1700s and early 1800s. Thus it was that by the Treaty of Ryswick in 1697, France acquired title to the French part of Santo Domingo — which they called Saint Domingue.

For administrative purposes, France divided the new colony into three provinces. North Province embraced the fertile Plaine du Nord and the colonial capital Cap Francais (later Cap Haitian). West Province comprised the central part of the French possession at the northern shore of the Haitian "notch" and the Cul de Sac. Its principal city, Port-au-Prince, was founded in 1749. The South Province included the rugged and almost inaccessible southern peninsula.

The colony's economic development was impressive. Magnificent plantations dotted the plains where thousands of slaves cultivated sugar, indigo, cotton and cocoa. Cap Francais became an opulent center of wealth and the arts, and was described as the "Paris of the Antilles." In European Paris the expression "as rich as a Creole" was proverbial.

Saint Domingue became renowned as the richest colonial possession in the New World, and for a time the value of its foreign trade was reputed to overshadow that of all 13 English colonies together on the North American mainland. On the eve of the French Revolution the annual value of Saint Domingue's foreign commerce was in excess of $78 million. More than 1,400 vessels and 11,000 seamen were employed in trade with Europe alone. Personal property was valued at almost $300 million.[1]

But the foundation of this wealth was tenuous. The 1788 census noted 28,000 whites, 22,000 *gens de couleur* (mulattoes and free blacks) and 405,000 slaves. The social fabric within each of the categories was rent by jealousy, suspicion and hatred. The native-born whites resented their French-born administrators. The mulattoes — owners of 10 percent of both land and the slaves by 1789 — were notoriously cruel taskmasters who drew the color line with paranoiac precision. The free blacks generally aspired to be slave owners, a status they seemed to covet as much as they had their own freedom. At the festering bottom of the social heap, the worst ignominy was to be a slave

1. International Bureau of the American Republics, "Haiti, a Handbook," Bulletin No. 62, U.S. Government Printing Office, 1893, pp. 87-88.

to another black.

The collapse of the fragile structure that was Saint Domingue in the closing of the 18th century came about not as a result of an internal failure of its underpinnings but from the winds of revolution sweeping through France — a storm felt in gale strength in the West Indies.

Six representatives of Saint Domingue's plantation aristocracy were seated at the French Estates General when it convened in 1789. If their hope was to secure a voice for their colonial constituency, they succeeded all too well. In France, emerging Jacobin politicians — egged on by rabid abolitionists called *Les Amis des Noirs* — spoke out in support of full political rights for the mulattoes and ultimately ended slavery in the colony. Saint Domingue was drenched with impassioned pleas for liberty and equality that stirred poor whites, mulattoes and blacks. The racial pot had reached such a raging boil that the third element in the French Revolutionary motto — Fraternity — disappeared in the steam.

On March 12, 1791 at Cap Francais, two young mulattoes — Ogé and Chavanne — were "broken on the wheel" on charges of agitation against the white aristocracy for citizenship rights promised by the Jacobins in France. The place where the young men perished at the hands of the executioner's rod became known as the Square of the Martyrs. Many more would perish in the gathering storm. Ogé and Chavanne sounded a tocsin soon heard loudly in the countryside surrounding the glittering Cap.

2

REVOLUTION AND INDEPENDENCE

So now, in the Autumn of 1791, looking from the sky-windows of Cap Francais, thick clouds of smoke girdle our horizon, smoke in the day, in the night fire; preceded by shrieking white women, by Terror and Rumor. Black demonized squadrons are massacring and harrowing with nameless cruelty.

They fight and fire 'from behind thickets and coverts,' for the Black man loves the bush; they rush to the attack, thousands strong, with brandished cutlasses and fusils, with caperings, shoutings and vociferation – which, if the White Volunteer Company stands firm, dwindle into staggerings, into quick gabblement, into panic flight at the first volley, perhaps before it.

Poor Ogé could be broken on the wheel; this fire-whirlwind too can be abated, driven up into the Mountains; but Saint Domingo is shaken ... writhing in long horrid death-throes, it is Black without remedy; and remains as African Haiti, a monition to the world.

— Thomas Carlyle, "The French Revolution."

The Jacobin winds from France had blown the elusive scent of freedom to Saint Domingue and during the summer of 1791 jungle drums thundered secret messages across the Plaine du Nord and up into the mountain fastnesses of Haiti. The revolt erupted on the night of Aug. 22, 1791 at Turpin Plantation, not far from Cap Francais. In the weeks that followed, bands of marauding blacks roamed the rich plain at will, killing, raping, looting, burning. The hated symbols of their past servitude — the bell towers that had summoned them to labor in the fields and cane mills — were razed along with the splendid homes of their erstwhile masters.

The marauders were eventually checked at the gates of Cap Francais itself.

But by then the surrounding countryside was in smoking ruins and white survivors of the holocaust were under siege in the cities and towns of the colony.

From an alarmed France, a commission was dispatched but its members — pledged to an anti-slavery policy and cheered on by the *Amis des Noirs* — only exacerbated the situation. The melancholy climax of this first phase of the Haitian Revolution was reached in June 1793 when Cap Francais suffered savage looting and burning. Refugees streamed across the border into Spanish Santo Domingo, and from every port in the French colony overcrowded ships sailed for the nearby islands, the United States and France, which was also in the grip of revolutionary terror.

The chief commissioner to Haiti — a man named Sothonax — remained remarkably unperturbed. He reported to France that "... it is with the real inhabitants of this country, the Africans, that we will save to France the possession of Saint Domingue."

On Aug. 28, 1793, apparently on their own authority, the commissioners proclaimed the formal abolition of slavery in the North Province, an act confirmed by the French government in February of the following year when it issued a decree abolishing slavery in all French colonies and declaring that all who were domiciled therein, regardless of color, were French citizens.

The Spanish in Santo Domingo and the British in Jamaica could hardly view these acts with indifference and, since the mother countries were both at war with France, the opportunity was seized to claim parts of Saint Domingue for themselves. Spanish forces, aided by a ragtag army of 600 blacks led by an ex-coachman named Toussaint, advanced into the French part of the island from the east. The British, welcomed by the remnants of the proprietor class in the South and West provinces, occupied Mole Saint Nicholas. By the spring of 1794 Spanish forces were in Port-au-Prince with a large and well-financed expedition. Toussaint, his force swollen to 4,000 men, won a dozen battles and achieved high rank in the Spanish Army.

At this moment, deemed the nadir for France in Saint Domingue, two unrelated events dramatically altered the military and political balance on the island. In April 1794, Toussaint with all his troops suddenly deserted the Spanish Army and joined the shattered French. The second event was a serious outbreak of fever among the men in the British expedition. The Spanish invasion of Saint Domingue collapsed almost overnight, and when, in 1795, peace came in Europe, Spain ceded all of the island of Santo Domingo to France in the Treaty of Basle. The British, bedeviled by disease and bitter guerrilla warfare, were forced back into their principal strongholds at Port-au-Prince and Mole Saint Nicholas.

The fortunes of war clearly favored France. But French power in Saint Domingue rested primarily on black warriors. By the fall of 1797, Toussaint-L'Ouverture ("The Opener") was commander in chief of the black armies and nominal custodian of French interests in the colony.

In October 1789 the British surrendered their last foothold on the island at Mole St. Nicholas. Like the Spanish they paid a heavy price for their adventure in Saint Domingue. Some 50,000 of their troops are thought to have perished in the campaign.

Toussaint next struck out at the dissident southern mulattoes, who were chafing under his "black" rule. He unleashed one of his lieutenants, Jean Jacques Dessalines ("the Tiger") on a brutal march through the South and West provinces. This was the first of many racial purges staining the pages of Haitian history. The mulattoes in check, Toussaint then sent another army to occupy the former Spanish part of the island.

By 1800 Toussaint — honored as "First of the Blacks" — ruled in the name of France all of Hispaniola, which had been devastated by nine years of war. The plantations, the mills, the irrigation ditches and the roads had fallen into decay throughout the island.

Meanwhile, across the sea, Napoleon Bonaparte brooded over the lost treasures of Saint Domingue. A temporary truce in Europe meant that Britain's sea power no longer was an obstacle to the movement of French troops and supplies across the Atlantic. Napoleon seized this opportunity and ordered the outfitting of an expedition entrusted to the command of his brother-in-law, Gen. Charles Emanuel Leclerc. In December 1801, more than 50 ships bearing 22,000 troops, many of them veterans of Napoleon's own campaigns, sailed for the West Indies.

July 1801, Toussaint-L'Ouverture, proclaimed the Governor-General, under the Constitution of the Republic of Haiti.

Leclerc, accompanied by his wife, Pauline, sailed with explicit instructions to secure Saint Domingue by a lightning campaign planned in detail by Napoleon himself. Pauline's brother also issued orders for the social reconstruction of the colony. "White woman who have prostituted themselves to Negroes, whatever their rank, shall be sent to Europe... Suffer no black above the rank of captain to remain on the island... Any person talking about the rights of those blacks who have shed so much white blood shall under some pretext or other be sent to France, whatever his rank in service."

On Feb. 3, 1802, the vanguard of the French fleet glided past the silent batteries of Fort Picolet and dropped anchor in the harbor of Cap Francais. The city was held by a small force under the command of Henri Christophe, one of Troussaint's trusted lieutenants. Prior to retreating, Christophe put the city to the torch so thoroughly that barely 60 buildings were standing when Napoleon's sister saw for the first time what was to have been her lavish New World home.

Despite severe fighting in the interior, Gen. Leclerc achieved all of his military objectives in but a few weeks. Toussaint and Dessalines surrendered on May 1 along with their troops. The "First of the Blacks" was spirited out of the country on the French frigate *La Creole* to France, where he died the following April in a dungeon at the fortress of Joux.

But in contradiction of Napoleon's orders, Dessalines was invited to join Christophe on the commissioned lists of the French Army when another amazing turnabout occurred. Malaria and yellow fever swept through the French ranks, and despite pleas of ever increasing urgency, Leclerc was unable to secure sufficient replacements from France even though Napoleon had already sent him 20,000 additional men. Laclerc thus had no choice but to take a desperate gamble: Perhaps he could retain control by playing the black generals off against each other. But that slim hope vanished when, late in the summer of 1802, word spread to Saint Domingue that by order of Bonaparte, slavery was re-instituted in the West Indies islands of Guadaloupe and Martinique.

By mid-October the remaining French on Saint Domingue were reduced to a precarious garrison existence in only six cities. On Nov. 28, Leclerc died of fever at the Cap, and Pauline returned to France. Leclerc's successor, Rochambeau, held on for another agonizing year while the noose gradually tightened. Port-au-Prince eventually fell to Dessalines, and a month later Rochambeau surrendered the remnants of his French Army at Cap Francais.

On Nov. 29, 1803, Rochambeau and 8,000 surviving soldiers sailed off from the Cap leaving behind the graves of nearly 45,000 brothers-in-arms who perished in Napoleon's vain attempt to restore Saint Domingue to French control. They also left behind a sizable white population that somehow managed to survive 12 bloody years of revolution.

Thus ended Napoleon's dream of a New World empire. His failure was an important factor — perhaps the main one — for his abrupt decision to sell the Louisiana Territory to the United States.

Rochambeau and his army were captured at sea by a British squadron and imprisoned for the duration of the war in Europe. Meanwhile, the uneasy whites left on Saint Domingue cowered in the unnatural silence of their cities.

3

THE BLACK REPUBLIC

We will write this act of Independence using a white man's skull for an inkwell, his skin as parchment, blood for ink and a bayonet as pen.

— Dessalines' secretary, at Gonaives, Jan. 1, 1804.

On Jan. 1, 1804, the day celebrated as Haitian Independence Day, the victorious black generals gathered at Gonaives and proclaimed Jean Jacques Dessalines "Governor-General of Haiti for life." According to legend, it was Dessalines himself who gave Haiti her name and her flag. With his own hand he is said to have torn the white stripe from the French tricolor and cast it underfoot. What remained would be the national ensign of the Haitian Republic. Appropriately enough, the blue stripe later was replaced by a black one.

African born, uneducated, ex-slave to an ex-slave, a man of ferocious instincts, Dessalines held in his heart pitiless hatred for whites. His first official act as governor-general was to decree death to all French still in the country. When subordinates proved reluctant to carry out this genocide, he himself led a murderous troop through those cities and towns still providing shelter for the condemned. The massacre began in the first days of January and continued into the spring. When it ended, the whites in Haiti were exterminated as a class. Not one in 10 escaped slaughter.

The news spread far beyond the Haitian frontier and was noted with particular concern and alarm in the United States, where refugees recounted tales that aroused fears of another black revolution.

The destruction of the white population in Haiti was accompanied by the devastation wrought on the land. No revolution, no war in modern history, ever so completely uprooted, demolished and eradicated the social, political and economic structure of a country. Haiti has never recovered to this day, and while the formal institution of slavery was abolished, personal and political

Haitian President Henri Christophe.

freedom for the Haitian masses did not attend the removal of the French. At best what occurred was the substitution of one form of despotism for another, neither one of which holds any special attractiveness in the pages of history.

The reign of Dessalines ended abruptly by an assassin's bullet on Oct. 17,

1806, at Pont Rouge on the outskirts of Port-au-Prince. Though his leadership was short in time, it was long in dubious achievement. In addition to the formal founding of the Republic and his "final solution of the French problem," Dessalines oversaw the writing of the first Haitian Constitution. Slavery was abolished forever and white men of whatever nationality were prohibited from acquiring property of any kind in Haiti. He attempted to restore the country's agricultural prosperity, an effort that failed.

Peasants were herded to work in the fields under a sharecropping scheme whereby they were to receive one-third of the harvest. Three groups claimed proprietorship over the ruined plantations. These were, first of all, the military chiefs; then the *ancien libre*, or pre-revolutionary free blacks and mulattoes who had held title to land; and lastly the mulatto sons of dispossessed white fathers.

With Dessalines dead, a military council named Henri Christophe as "Provisional Chief of the Nation." At the same time, however, the council called for a constitutional assembly and the formal election of a new head of state. This assembly, the first legislative body created in Haiti, met in Port-au-Prince on Dec. 18, 1806. One of the leaders was Alexander Pétion, a mulatto educated in Paris who had returned to Haiti as an officer in Leclerc's army. Christophe was quick to sense in Pétion the chief rival for power. The political struggle between the two ultimately degenerated into a fitful civil war. After several skirmishes during which Christophe managed to seize a large part of Dessalines' treasure, He withdrew his army to the north. Haiti was in effect split into two parts, with Pétion in control of South and West provinces, and Christophe in command of the North.

Until March 1811, Christophe governed his part of the country as "President of the State of Haiti," but in that month he published the "*Loi Constitutionelle du Conseil d'Etat qui établit le royauté a Haiti.*" On June 2, the chiefs of the North gathered at the old French colonial capital to witness the coronation of Christophe. The Archbishop of Haiti, recently elevated to that post by Christophe, crowned the new monarch Henri I. Nobility was established with four princes, eight dukes, 22 counts, 37 barons and 40 chevaliers, among whom were a Count of Limonade and a Duke of Marmalade (titles taken from districts in North Province facetiously named by the French). Cap Francais, recently Cap Haitian, was renamed Cap Henri.

By the king's order in the Code Henri, every adult man and woman in the new kingdom was required to work "from daylight to 8 o'clock, then one hour off for breakfast on the spot; from 9 to 12, then two hours off; and from 2 p.m. until nightfall."[1] The soldiers of the king enforced this regimen and an almost instant prosperity returned to the North.

Creating a circulating currency was accomplished by declaring all ripe gourds property of the state. These were collected by the Crown and parceled out as wages. The gourd remains useful to Haitian peasants in the manufacture of pots, tools, containers, musical instruments, etc. Gourds were paid for work performed on state lands, or in exchange for homegrown crops, principally coffee. To this day the unit of Haitian currency is called a *gourde*, though it has long since ceased to be an actual one.

Construction of the Citadelle la Ferrière, Haiti's greatest architectural monument was begun by order of Dessalines in 1804 and largely completed by Christophe. In shape, it is an 18th century warship built with incredible labor on a mountain peak called *La Bonnet a l'Eveque* (the Bishop's Hat) 3,000 feet above sea level and 20 miles inland from Cap Haitian. The prow of this great fortress points to magnetic north, and its walls, 20 to 30 feet thick, range in height from 80 to 130 feet. Three-hundred-and-sixty-five cannons, some of them 11 feet in length — trophies captured from the French, the British and the Spanish — were dragged up the mountain and placed in the fortress along with immense quantities of shot and powder. The Citadel's construction continued throughout Christophe's reign and, as attested by events immediately following his death, was still in progress when his kingdom dissolved. Christophe seemed obsessed with the necessity of completing the work, driving his people mercilessly. He himself is said to have labored on the walls as a mason, sometimes by moonlight.

In 1820 he suffered a stroke that left him paralyzed from the waist down. Deserted by his servants, the nobility and the army, he committed suicide at his palace, Sans Souci, at the foot of the mountain. Allegedly, he shot himself

1 J.W. Vandercook, "Black Majesty: The Life of Christophe." New York, Harper and Bros., 1928. p 134.

Citadelle la Ferrière, Haiti's greatest architectural monument, was begun by order of Dessalines in 1804 and largely completed by Christophe.

with a silver bullet. Somehow, his wife and two daughters managed to drag his body up the mountain to the Citadel, where it was placed in a pit of newly mixed mason's lime. The spot is marked by a weathered stone inscribed simply, *Christophe — L'Homme* (The Man).

In 1818 Christophe's great rival, Pétion, who died in office of natural causes (one of the few Haitian presidents ever to enjoy that distinction), was succeeded by the mulatto commander of the palace guard, Jean Pierre Boyer. Boyer used his position to secure unanimous election to the presidency. He began his 25-year career as Chief of State on March 20, 1818. Christophe's suicide two years later enabled Boyer to reunite the country. The only disappointment Boyer felt in a bloodless campaign in the north was the discovery that in King Henri's treasury there was only 10 million francs, instead of the hoped for 250 million.

In 1821, the Spanish part of the island revolted from Spain, which had reclaimed it during the troubles following the death of Dessalines. With little persuasion it attached itself once more to the Republic of Haiti. Thus, the whole island was again governed from Port-au-Prince.

Boyer's rule, the longest in Haitian history, was marked by unsuccessful attempts to restore the island's prosperity. The great estates of colonial times were divided into small parcels, but land reform seemed only to hasten separation of the people into two rigid social classes — the peasants and the elite. The former subsisted by sharecropping small plots or by harvesting upland coffee, which then grew wild. The latter gravitated to the cities, where many lived on the labor of others by imposing and collecting export duties on their cash crops.

Prior to Pétion's death, a more liberal constitution was adopted and authority divided between executive, legislative and judicial branches of the government. A court system was established and a large bureaucracy was formed to satisfy the political aspirations of the elites, mostly mulatto, who openly harkened back to and embraced the old French-Catholic culture.

From the very beginning of elite rule, certain practices crept into the treasury service and firmly rooted, which seriously hampered government operations. One of the worst of these concerned the system adopted for paying public employees, including those in the army and navy. Payments were customarily made by claims on the treasury called *feuilles*, which were often left unredeemed for long periods and particularly during *la saison morte,* the time between crops when customs receipts dropped and there was insufficient funds in the treasury to cash them.

Speculators routinely purchased feuilles at large discounts, and by exerting influence at high levels of government, got preferential treatment. Often the government itself purchased feuilles at discount. Graft and dishonesty were invited by the administrative procedures that were adopted and eventually became ingrained in the ethics of Haitian public service.

An attempt was made to revitalize the economy by encouraging immigration of people of African or Indian origin, particularly those from the United States. In a four-year period some 13,000 freedmen were enticed to Haiti with promises of free land and other entitlements. The experiment ended in dismal failure. Most of the immigrants left Haiti at the first opportunity or abandoned the land given them and returned to the United States. Many said that discrimination in Haiti made their lives little better than slavery. Few were able to adapt to the customs, religion, language and laws in Haiti.

In 1822 Boyer appealed to the United States for diplomatic recognition, but President Monroe declined, citing France's decision to deny it. He also noted that Haiti was not open to fair trade with America, and that Haiti's constitution still embodied anti-white articles.

French recognition was secured in 1825, but at a price. Under the guns of 14 warships anchored in the harbor of Port-au-Prince, Haiti agreed to pay an indemnity of 115 million francs as compensation to former French colonists. A second concession was that French trade with the island would be assessed at only half of the duty charged other countries.

The financial drain to meet even partial payment of the indemnity was so great that in 1827 Boyer was forced to issue the first Haitian paper currency. His success with it was not nearly as great as had been with Christophe's gourdes. There was no possibility that Haiti's vast foreign obligation would ever be discharged, and the nation's inevitable default brought on what was the first of many such crises in its foreign relations.

Haiti's economic ills, growing discontent among the masses and dissatisfaction in the army led to a revolt in January 1843. In March of that year, Boyer resigned from office and found refuge on a British warship. He died in Paris in 1850.

Following Boyer's resignation, another period of political and racial turmoil shook the island. The peasant blacks, promised impossible advantages by those who had organized the overthrow of the president, felt deceived. Vicious class warfare broke out and in the general confusion the Dominicans, in 1844, reasserted their independence. The sudden death of Boyer's nominal successor, Jean-Baptiste Riché, on Feb. 27, 1847, caught the mulatto elites who controlled the Haitian Senate unprepared.

To placate the blacks, they agreed to select as president — some said by drawing his name from a hat — the 60-year-old illiterate head of the palace guard, Faustin Soulouque. He was thought to be both naive and agreeable, a most willing instrument through which the mulattoes could exercise political control while satisfying the demands of the masses. Within two years, however, Soulouque learned enough about the exercise of power to overthrow his patrons and carry out in April 1848 a massacre of mulattoes in Port-au-Prince. His hold on the government thus secure, he turned his interest on the first of a series of disastrous military campaigns designed to re-annex the Dominican Republic.

On Aug. 26, 1849, at the "urging" of his officers, Soulouque assumed the tile of Emperor Faustin I. The coronation was an extravagant affair that took place in a scene of great public jubilation at Port-au-Prince. The nobility created by Faustin I dwarfed even that of Christophe. The emperor adopted as his motto, "I am the State and my will is law."

In spite of an almost paralyzed economy, Faustin pressed his prolonged and costly invasions of the Dominican Republic. Both countries were pushed to the brink of financial ruin by this war, which was finally ended by the diplomatic intervention of the United States, Britain and France. His hold on the army weakened by the Dominican failures, and with the country in a severe and worsening financial depression, Faustin was deposed by mulatto general Fabré Geffrard in January 1859.

The eight years of Geffrard's rule cast a relatively bright light on the political darkness of 19th century Haiti. In 1860 a concordat was signed with the Holy See and, in exchange for state financial support, Rome assumed responsibility for the propagation of the Catholic faith in Haiti. Clergy were dispatched to bring the island's faithful back to orthodoxy.

On June 5, 1862, Abraham Lincoln extended formal diplomatic recognition to the second oldest "republic" in the Western Hemisphere. In Boston, abolitionist societies celebrated the act as "a recognition of the Colored Man, not merely of Hayti."[2]

In 1864 a Treaty of Amity, Commerce and Navigation with the United States affirmed Haiti's most-favored nation treatment, a status enjoyed on an informal basis since 1850. The treaty further stipulated that Americans resident in Haiti and Haitians resident in the United States would not be required to pay "any contributions whatever, higher or other, than those that are or may be paid by native citizens."

The years of Geffrard's rule were also marked by the final efforts to foster

2 Quoted in Ludwel L. Montague, "Haiti and the United States, 1714-1938," Russell and Russell, New York, 1966, p. 86.

black colonization of Haiti from the United States. In 1860, an emigration bureau was opened in Boston under the direction of James Redpath, a militant abolitionist and associate of John Brown. Colorful and attractive brochures were printed and circulated. Prospective settlers were promised free passage, land, temporary housing, subsistence and exemption from military service. By November 1861 approximately 1,200 U.S. blacks had been recruited for Haiti. But distressing stories of privation, disease and ill treatment of the pioneers filtered back to Boston, and by the end of 1861 public meetings were held to denounce the colonization scheme.

In all, Redpath was responsible for sending about 1,600 American blacks to Haiti, but in 1862, only 200 could be accounted for. Some, it was assumed, managed to return to the United States.

Despite this dismal experience, the government in Washington embarked on a colonization scheme to alleviate the growing problem of dependent slaves freed by the Civil War. On Dec. 31, 1862, President Lincoln contracted with a private organization to settle 5,000 of them at Ile-a-Vache in Haiti. This also ended in debacle, and shortly thereafter the Haitian government lost interest in such projects. Only 431 freedmen were transported to Haiti by the federal contractors, and most returned to the United States.

The repeated failure of the attempted colonization pointed out a distressing fact. The Haitian who had won his freedom in a bloody revolt against France rather openly despised his American brother who, in Haitian eyes, had been given his freedom. This prejudice was deeply ingrained, and in the years ahead cultured Haitians would look with disfavor on the assignment of American blacks as representatives or employees of the United States in Haiti.

Revolutionary activity at Cap Haitian in 1866 led to the British consulate there being violated. In reprisal, British warships shelled the city. In September of that year, the arsenal in Port-au-Prince was blown up, with the loss of many lives and great destruction of property. When in the following year the presidential guard revolted and fired on the palace, Geffrard took the route of his predecessors. On March 13, 1867, he sailed for Jamaica.

Geffrard's successor, Sylvain Salnave, tried to repeat Soulouque's maneuver, and encouraged the peasants to rise against the mulatto elite. A protracted period of civil strife ensued. In 1870 the presidential palace, converted by Salnave into a powder magazine, was struck by a shot fired from a captured government vessel in the harbor. The palace was destroyed by explosion and fire.

President Salnave escaped to Santo Domingo, where he was arrested and returned to Port-au-Prince. Bound to a stake amid the rubble of the palace, he was executed by a firing squad. An editorial that appeared April 6, 1871 in the *Gazette du Peuple*, published in Port-au-Prince, is revealing:

"For sixty-eight years, what have we done? Nothing, or almost nothing. All our constitutions are defective, all our laws are incomplete, our customs houses are badly administered, our navy is detestable, our finances are rotten to the core, our police are badly organized, our army is in a pitiable state, the legislative power is not understood and never will be, the primary elections are neglected and our people feel not their importance, nearly all our public edifices are in ruins, the public education is almost entirely abandoned...."

Following Salnave's execution a succession of "presidents" entered and/or left office by unconstitutional means. Nissage-Saget (1870-1874), Michel Domingue (1874-1876), Boisrond Canal (1876-1879) and Lysius Salomon (1879-1888) made brief appearances on the disordered pages of Haitian history.

The presidency of Florvil Hyppolite (1889-1896) gave the country a brief respite from its long political misery. Foreign investment was encouraged, and a bond issue, at ruinous cost, funded French war claims. Some schools were established and a body of law was enacted. Public buildings were constructed, most notably the famous "iron markets" that still stand in Haiti's principal cities. There was even a serious effort to limit graft and improve the efficiency of government.

But representative government was not functioning, and indeed had never really done so, as Haiti prepared to enter the 20th century.

4

THE WIDENING AMERICAN CONCERN

As far as can be seen, the time will come when stable governments for the American tropical States must be assured by the now existing powerful and stable States of America or Europe. The geographical position of those States, the climatic conditions, make it plain at once that sea power will there, even more than in the case of Turkey, determine what foreign State will predominate — if not by actual possession, by its influence over the native governments. The geographical position of the United States and her power give her an undeniable advantage; but that advantage will not avail if there is a great inferiority of organized brute-force, which still remains the last argument of republics as of kings.

— Alfred Thayer Mahan, "The Influence of Sea Power Upon History 1660-1783."

It is perhaps well at this time to depart for a moment from the chronological account of Haitian history and review the several aspects of American interest in the Black Republic. The picture is somewhat complex, for there was ever a wide variance of views and conflicting concerns among Americans with economic or emotional ties to Haiti.

American policy was thus forged in the heat of opposing forces, those that seldom if ever seemed to add rationally as vectors on a force diagram. It is just this aspect of policy formulation that makes its study so intriguing, for seemingly insignificant acts and events have from time to time become governing, and have directed American policy in unpredictable ways. Thus it was, as will be seen, American intervention in Haiti came after what might be considered compelling reasons for it — the desire for a Caribbean naval base and a need to forestall European intervention. Those reasons ceased to exist as a result

U.S. President Woodrow Wilson.

of the American war with Spain and the involvement of the European powers fighting among themselves. And, paradoxically, the intervention came not under President Theodore Roosevelt, but President Woodrow Wilson.

It is as if a peculiar inertia shaped the affairs of men. This inertia was like a great weight attached to a strong but elastic line tugged at for years by leaders intent on moving it in their own direction. When it finally begins to move, the elasticity of its tether carries it beyond the position of its first movers who may, in fact, already have surrendered their hold on the line to others of opposing inclination. The process may then be seen to begin again in reverse.

This swing effect in American foreign policy is as observable as the rise and fall of ocean tides, but its regularity and predictability are by no means so precise. The forces of change and inertia are ordinarily so nicely balanced that a minuscule weight applied at a critical time may achieve an effect that is grossly disproportionate and even, to an uncritical eye, irrational.

American interest in Haiti predated the independence of both its republics. The sugar islands of the West Indies, of which Hispaniola was by far the richest, were an integral link in the chain of trade on which rested the prosperity of New England merchants. Slaves and fish products were traded in the islands for cheap molasses that was converted into rum in New England towns and sold to acquire more slaves and expand the fisheries. British interference with this trade (manifested by the several Molasses Acts) can be seen as a prime irritant leading to the American Revolution.

Haitian and American independence did not alter the basic commercial relationship between Haiti and the New England states, although the Black Republic no longer provided them a market for African slaves. The Haitian revolution did, however, cause considerable alarm and foreboding in America's southern states. Henceforth, a profound dichotomy of interests concerning Haiti developed along geographical lines in the United States. In general, the North urged diplomatic recognition and expanded economic ties with Haiti because it seemed it would be good for business. But the South routinely blocked recognition and cared little for the trade interests of the North. It is significant that recognition of Haiti came only after the South's secession from the Union.

Naval developments, the Louisiana Purchase and concern over effective control of an isthmian route to California influenced American interests in Haiti. American naval power, a late arrival on the Caribbean scene, found the best sites for naval bases already taken by European powers, with Samana Bay in the Dominican Republic, St. Thomas in the Danish West Indies and

Mole St. Nicholas in Haiti the only attractive sites thought to be available.

The decline of sail and the emergence of steam as a means of naval propulsion made such a base in the Caribbean even more vital in the eyes of naval strategists, for the new propulsion system "shortened the legs" of warships and made necessary the pre-positioning of large stockpiles of coal in the areas where such vessels would be required to operate.

Pressures for Caribbean bases and coaling stations peaked during and shortly after the American Civil War. The decline of the great Union fleet after that war and the completion of the first transcontinental railroad in 1869 (which lessened for a time American interest in the isthmian route to California) cast an eclipse on the project of Caribbean bases. This interest, despite aborted attempts to annex the Virgin Islands and the Dominican Republic, did not revive in earnest until American interest turned again toward an isthmian canal and the rebuilding of the U.S. Navy.

How much did the philosophy of "manifest destiny" shape American policy in Haiti during the latter part of the 19th century? Throughout U.S history there has been an unfortunate fascination with a presumed mission to civilize backward nations, but not so for Haiti. Apparently there was no serious consideration given to the annexation of Haiti, although President Andrew Johnson suggested it in his message to Congress on Dec. 9, 1868.

The political and racial realities, to say nothing of the experience of France on the island, dampened annexationist spirits in the United States. To many, the Haitian problem seemed insurmountable because of racial strife in America. The last serious negotiations for a base at Mole St. Nicholas was in 1889-1890 when, in spite of considerable U.S. assistance given Haitian President Hyppolite in his climb to office, the negotiations failed, probably because of his nation's constitutional bar to white ownership of property.

If the fire of manifest destiny dimmed in Haiti, that other beacon of U.S. policy in the Caribbean, the Monroe Doctrine, most definitely did not. Several incidents already cited illustrate that the guns of foreign warships were not slow to enforce what was considered vital state interests. The threat to the Monroe Doctrine lay in the chance that a casual bombardment or the landing of troops might grow into full-scale and permanent foreign occupation.

Gradually the feeling grew in governing circles that if military demonstrations were necessary to keep Haitian affairs in order, they should be American, not European. This theory foreshadowed Theodore Roosevelt's corollary to the Monroe Doctrine.

Visits to Haiti by U.S. naval vessels became more frequent and, in 1888, Marines were landed in Port-au-Prince to quell a disturbance that seemed to threaten the safety of foreign nationals. The European powers did not accept exclusive American protection of their interests, however, and their own warships continued to call at Haitian ports. A glaring example of foreign military pressure, and one with somewhat humorous overtones, occurred in 1887.

A German national named Emil Luders, the operator of a livery stable in Port-au-Prince, attempted to prevent the arrest of one of his Haitian employees by claiming extra-territoriality for his stable. He was arrested on a charge of assault and battery growing out of the affair, sentenced to a year in prison and fined $500. Luders, however, was a veteran of the German army, having served in the Kaiser's regiment. A German imperial letter was dispatched to the Haitian president, Simon Sam, demanding the release of the distinguished livery stable operator, the punishment of those responsible for the arrest, the removal of the judges who had passed sentence and payment of a large indemnity.

The Haitian government appealed to the United States for aid, citing the Monroe Doctrine, but none was forthcoming. When the Kaiser sent two cruisers to enforce his demands, the American Legation in Port-au-Prince felt that the United States had made a mistake, and was losing influence in Haiti — a position adhered to by prominent Americans including Theodore Roosevelt and Henry Cabot Lodge.

The Spanish American War and its immediate aftermath saw a dramatic increase in U.S. naval power and a growing tendency among certain Americans, made heady by the easy victory over Spain, to consider the Caribbean an American *mare nostrum*. The acquisition of naval bases in Puerto Rico and Cuba (Guantanamo Bay) made others in the Dominican Republic and Haiti unnecessary, but American admirals still expressed concern lest another sea power — and in this period they most often meant Germany — should establish itself there. The Navy's policy towards Haiti thus changed from one advocating acquisition of bases on the island to one of denying them to potential enemies.

The broader aspects of American policy in the Caribbean were stated clearly by President Theodore Roosevelt in his message to Congress on Dec. 6, 1904. Henceforth it would be the resolve of the United States to exercise an international police power in the Western Hemisphere should chronic wrongdoing on the part of delinquent states require it. Gradually, the arrival of a U.S. warship in a troubled Haitian port became the accepted signal for a changing of the guard.

Meanwhile, its finances a shambles and its politics impossible, the Haitian state careened along its dangerous way.

5

THE WINDING PATH TO ANARCHY

However, it is visible that God protects our little corner of the earth; for notwithstanding the faults of its children, the country is still unimpaired, and will so pass into the hands of our children's children.... And thus if all citizens give me their aid, we would easily avoid the misfortune that has fallen on Dominica.

— Haitian President Nord Alexis, Feb. 5, 1905.

At 7:25 on the morning of April 11, 1900, a large meteor of great brightness crossed the sky over Port-au-Prince trailing a long train of nebulous matter. About three minutes after its passage, a loud explosion shook the city. The phenomena caused considerable anxiety among the Haitian populace, who looked upon it as foretelling dire disaster in the Republic. A sign from heaven was scarcely needed, so precarious had become the condition of the Haitian state. Yet another financial crisis was throttling the life of the economy, and grave constitutional questions were being aired concerning President Simon Sam's term of office.

The Haitian external and internal bonded debt was ordinarily secured by pledges against customs receipts. These were paid into the National Bank of Haiti, a French corporation chartered on April 1, 1880 during the administration of President Saloman. The signer of this concession for Haiti was the president of the National Assembly who had a peculiar and, judging from later circumstances, humorous name — Innocent Coco.

The 1900 crisis was caused by a simple fact: An inadequate part of the customs receipt had found its way into the state treasury — that is, the Bank. The Haitian customs service, needless to say, was notoriously corrupt. By a remarkable feat of legerdemain, the public debt was refinanced and consoli-

Port-au-Prince public market.

dated, thus further mortgaging anticipated state revenue. Two classes of bonds, payable in 14 years, were issued. The first, known as *Les Grandes Consolidations*, bore interest at 12 percent per annum. The second, *Les Petites Consolidations*, yielded only 6 percent. It was stipulated that both interest and principal were payable in American gold. The consolidated debt was approved by a unanimous vote of the Haitian Senate and the House of Delegates. That vote itself should have been cause for suspicion. The consolidation and the role of the National Bank merit further attention, but first it will be necessary to examine the political situation that existed at the time.

Articles 90 and 93 of the Haitian Constitution concerned the election of the President:

"Article 90. The President of the Republic is elected for seven years. He will enter upon the duties of his office on the 15th of May. Only after an interval of seven years [may he succeed himself]."

"Article 93. In case of the death, resignation or forfeiture of the President, the one who replaces him is named for seven years and his functions end always on the 15th of May, even though at that time the seven years of his powers are not finished. During the vacancy the executive power will be exercised by the Secretaries of State in Council assembled, and under their responsibility."

The joker in the case of President Simon Sam — who was elected by the National Chambers on April 1, 1896, following the sudden death of President Hyppolite — was that at the time of President Sam's election, a law was enacted requiring him to enter upon the duties of the President at once and remain in office until May 15, 1903.

The Haitian Constitution did not authorize judicial review of acts of the legislature, such as that exercised by the U.S. Supreme Court. Thus there was a body of opinion in Haiti (the "Ins") holding that laws enacted by the legislature and approved by the executive were perforce constitutional. Another body of opinion (the "Outs") thought otherwise.

According to Article 93 of the Constitution, President Sam's term in office would expire on May 15, 1902, but by the 1896 law his term would not expire until a year later. The electoral system in Haiti, patterned after the French, further aggravated the situation. The Chamber of Deputies was elected by popular vote to serve for a term of two years. The chamber was responsible for the election of the upper house, the Senate, whose members served six-year terms. When a President was to be elected, both houses met in joint session as the National Assembly and the election took place at the congressional session just preceding the expiration of the presidential term. By tradition there were many candidates, for not only honor and a legal salary of $24,000 in gold awaited the successful candidate, but almost unlimited opportunity to pillage the state treasury.

Not surprisingly, the electors were frequently wooed with promises of material advancement in addition to eloquent appeals for patriotism. If this judgment of Haitian political morality seems harsh, it is nevertheless believed to be accurate.

Adding to the comic opera overtones in Port-au-Prince at this time was the maneuvering of the great powers for commercial advantage. France, the year before, had signed a treaty with Haiti that reduced tariffs on French imports.

Ceremonial honors were also heaped on high officials in the Haitian government. Haiti's minister of finance and its minister in Paris received the Grand Cross of the Legion of Honor, and on President Sam was conferred the *Grand Croix de l'Ordre de l'Etoile Noire.*

Paris financial journals commented favorably on the care Haiti's responsible officials took to meet the country's financial obligations. In addition, the U.S. minister in Port-au-Prince, William F. Powell, never ceased to act the role of a one-man chamber of commerce. He consistently touted U.S. investments in Haiti, particularly in banking and railroads. He entertained lavishly. A ball given by the Powells in honor of President McKinley's second inauguration earned special praise in the Port-au-Prince press. Guests entering the U.S. Legation were greeted with a huge eagle with outspread wings fashioned in artificial flowers. It held in its beak a ribbon with the inscription, *"E Pluribus Unum."*

The German community in Haiti, meanwhile, had taken the much more practical step of subscribing substantial funds toward the expenses of building new ships to be added to the German navy. By coincidence, shortly after the announcement of the German subscription, an Italian cruiser paid a visit to Port-au-Prince and pressed, successfully, a claim against the Haitian government.

Apparently President Sam planned to serve at least until May 15, 1903. Certain irregularities in the January 1902 legislative elections, however, created such uproar that it became obvious he would have to step down sooner. It was said that in many places in the country, those who were allowed to vote had to promise to cast their ballots for government-approved candidates. In other cases, where opposition candidates received a majority of the vote, their elections were set aside and others were "elected" in their places. Those who protested were arrested. As in all Haitian elections, great use was made of the army. Troops were marched to the polls and their vote cast as a group.

The National Assembly delayed a vote for a new president until May 12, 1902, when apparent agreement was reached on President Sam's choice of a successor, Gen. Cincinnatus Leconte, who was thought to have the backing of the German community. A rumor circulated that Gen. Leconte had promised Germany a coaling station at Mole St. Nicholas. An angry mob forced its way into the legislative chambers just prior to the ballot. Shots were fired and the disturbance spread into the streets. At least 100 people are thought to have lost their lives.

At 6 a.m. the next morning, the American minister received a summons from the palace. Powell was asked, in his capacity as dean of the diplomatic corps, to grant protection to President Sam and his cabinet. Powell called his colleagues to a hasty meeting at the U.S. Legation. At 11 a.m. the diplomats as a body escorted the Haitian president and senior officials of the government to the harbor where they embarked on a steamer. Sam rode to the steamer in Powell's carriage.

Powell told Washington, "the President, General Sam, retires from office moderately well fixed for his future, having been able to secure and place to his credit in France about 12,500,000 francs, or $2,500,000."[1]

As was customary in Haitian revolutions, committees of public safety were established in Port-au-Prince and all other major cities and towns in the country to continue the essential functions of government and, not coincidentally, to exert influence over the choice of a provisional government that would be formed pending new elections. Leading candidates in the changed political circumstances busied themselves raising armies.

There were threats to march on the capital. Gen. Firmin, commander of the Army of the North, quickly emerged as the strongest candidate to succeed to the presidency. Despite instructions to remain strictly neutral, Powell persuaded his French, German and British colleagues to join him in sending a letter to Firmin urging him to keep his army clear of the capital. The diplomats also urged their home governments to send warships to Port-au-Prince during the present unrest.

Powell had reason to be grateful for such a visit by USS *Topeka*. While escorting the particularly hated ex-minister of interior and police Tancrede Auguste to the harbor, a mob gathered and had to be dispersed by sailors from the American cruiser.

Late in May 1902, representatives from the various committees of public

1 U.S. National Archives. Powell to Dept. of State, No. 1092, May 17, 1902.

safety met in Port-au-Prince. A provisional government was established with Boisrond Canal, an elderly ex-president, at its head. Firmin was not represented. In July another election was held, but at almost the same time a vessel of the Hamburg-America line arrived from New York with a cargo of 1,500 rifles and coal for Firmin in Gonaives. The rifles were carefully packed in barrels labeled "pork." The coal was for the antique Haitian gunboat *Crete à Pierrot*, with mulatto Adm. Hamerton Killick aboard. He and his ship joined forces with Firmin, and the civil war was promptly renewed.

The expansion of Haiti's domestic quarrels to the high seas raised questions concerning international law, in particular that about an effective blockade. Killick proclaimed a blockade of certain Haitian ports, though he and his gunboat clearly had no means of enforcing it. The provisional government responded by declaring ports controlled by Firmin's army also in a state of blockade. The United States, which 41 years earlier had seen its "paper blockade" of certain Confederate ports properly laughed at, this time adhered strictly to international law.

U.S. naval officers were directed to give protection to American and unprotected foreign vessels if it were determined that declared blockades were not effective. Given the state of the contending Haitian navies, such determination was almost a foregone conclusion. Nevertheless, Killick in the *Crete à Pierrot* prevented the American merchant ship *Paloma* from entering Cap Haitian, and commercial interests went into an uproar. On Sept. 2, 1902, Killick made a fatal error when he stopped, searched and removed goods from a vessel of German registry.

The German chargé d'affaires immediately cabled his government, and four days later the German gunboat *Panther* found the *Crete à Pierrot* anchored in the harbor at Gonaives. An ultimatum was delivered, at the expiration of which the *Panther* opened fire. Killick ordered three kegs of powder, some large cartridges and a can of kerosene to be placed inside his cabin, with a powder trail laid to the quarterdeck. He then ordered his officers and men ashore. The last thing he was seen to do was to fire the powder train with a lighted cigar. A terrific explosion tore him and his ship apart.

The *Panther* steamed out of the harbor ignoring the pleas of frightened

German residents. An angry mob filled the streets and shouted, "Kill the Germans! Kill the whites!" [2]

Killick's charred body was found floating in the harbor the next day. It was taken ashore and buried with honors. Thus did Haiti acquire her first naval hero, even though he commanded forces dedicated to the overthrow of the national government.

Though his naval arm was now destroyed, Firmin continued to press his struggle against the provisional government, which was having difficulty fielding an army. Powell reported that "in the mountain districts (recruits) are hunted like wild animals and driven into the cities like a drove of cattle with their legs tied together with rope, sufficiently long to enable them to walk, and their arms tied behind them."[3]

Powell contended in a series of dispatches that prolongation of the Haitian civil war was primarily due to foreign commercial houses making money out of both parties in the conflict. Large loans guaranteed by a percentage of the already heavily pledged customs revenue were made to both the provisional government and to Firmin. Arms and munitions at double their real value were similarly supplied. Those providing money and arms knew that their interests were secured regardless of which side won, for as in the past their home governments would apply diplomatic pressure backed by the threat of naval bombardment to whatever Haitian authority secured control. Under these circumstances foreign interests could not lose. An interesting illustration of this game of playing both ends against the Haitian middle is that it was a German ship of the same line as the one stopped by Killick that supplied the coal needed by the *Crete à Pierrot.*

Powell's solution to the Haitian dilemma was the displacement of foreign commercial interests by American ones. He was thus an early and most persistent advocate of "dollar diplomacy," a circumstance that lead to his own downfall.

The provisional government finally found a military leader of some ability

2 Powell to Dept. of State, No. 1158, Sept. 13, 1902.

3 Powell to Dept. of State, No. 1171, Sept. 22, 1902.

in the aged Gen. Nord Alexis. Both St. Marc and Gonaives were taken, and Firmin and his senior followers forced to leave Haiti.

On Dec. 14, 1902, Gen. Nord Alexis, a vigorous 85, entered Port-au-Prince at the head of an army of about 5,000 men. There were a few minor skirmishes, and around 15 people were killed. The political situation quickly stabilized, and the army proclaimed its general President of Haiti, an act ratified four days later by the National Assembly.

As might have been expected, the new president found state finances in utter chaos. The public debt had skyrocketed to nearly $29 million. Emergency measures were taken to double the duty on imports, to tax government bonds (originally issued as tax-free instruments) and to drastically increase the printing of paper money. A special commission was named to look into the recent consolidation of the public debt. Grumbling in the House of Delegates was silenced by forcibly dissolving that body. When the minister of finance resigned, no one volunteered to replace him.

Late in June 1903 the special commission — comprised of two judges of the Supreme Court, a top official of the National Bank and two businessmen — made an interim report to the president. They charged the bank with being wholly responsible for the issue of $2 million in fraudulent bonds, and that the bank itself had shared in the proceeds of that issue. The French director of the bank, Mr. J. de la Myre, denied the charge. Then he told President Nord Alexis the government had overdrawn its account and funds would not be advanced to pay the army. The president was quoted as telling his officers there was money in the bank and that they should go and get it.[4]

The more the consolidation scandal was investigated, the worse it seemed. Orders were issued for the arrest of all members of ex-President Simon Sam's cabinet, and one of them, unlucky or foolhardy enough to have delayed his exodus flight, was arrested and interrogated by the special commission. He testified that the fraudulent issue of bonds had been made and distributed, that he had strongly opposed the measure when it was introduced at a cabinet meeting and that he had been outvoted.

4 Powell to Dept. of State, No. 1316, July 2, 1903.

The investigations continued, and by October 1903 the most glaring frauds, amounting to nearly $3 million, had been discovered. In virtually all of the frauds the National Bank of Haiti had profited equally with the late administration. The bank, on its part, claimed that it was not responsible because it had acted at the direction of the Haitian president and the minister of finance.

In November, President Nord Alexis ordered the arrest of all senior officers of the National Bank, including Director de la Myer. Another diplomatic crisis threatened, for most of the arrested officials were French under the protection of the French minister.

At year's end the government's position was desperate. Public employees had not been paid for months, and while in the past some money had been found to disperse for the holidays, this year there was none.

Only the army was paid, the government borrowing at high interest from smaller commercial houses in exchange for customs due bills. No money was on hand to make preparations for Haiti's centennial, talked about for years and due to be celebrated on Jan. 1, 1904.

To keep the opposition in check, the government resorted to terror, mass arrests and public executions. Foreign legations were filled with refugees. U.S. Minister Powell reported that many of those fleeing the police entered his grounds by scaling a 20-foot wall. In mid-January, the German minister threatened to have marines landed to quell disturbances at Port-au-Prince.

The distress of the Haitian people worsened as the year progressed. A sharp fall in the foreign exchange brought huge increases in the price of imported goods. Much of Haiti's retail trade was then in the hands of "Syrian" merchants (of all nationalities except German). The government launched a vicious campaign against them, stirring up long-smoldering racial hatreds. In March 1905, all who were not naturalized or citizens of some other country were ordered out of the country by April 1.

Shopkeepers made frantic efforts to sell their goods and close their stores. Many had large families and were terrified by freely circulating stories about machetes being sharpened by soldiers and the lower classes. Powell requested the presence of a naval vessel in Port-au-Prince, fearing an outbreak of looting and arson. Traffic boomed in counterfeit naturalization papers.

On April 1, 1905, those so-called Syrians who could not secure the protection of another state were forced out of Haiti. Their departure was almost immediately followed by another round of spiraling inflation. Powell reported that, "There are many families here among the middle class that are in a state of almost semi-starvation.... Many families are parting with their jewelry and articles of wearing apparel to buy food."[5]

Meanwhile, the trial had begun of those connected to the consolidation scandal. Powell thought the Haitian government had placed itself in a compromising position when its minister of foreign relations informed the French minister that bank officials had to be tried and condemned for political reasons. The French minister lost no time in communicating that to his government and to the press. Correspondents from leading Paris newspapers were sent to Port-au-Prince to cover the trial.

The Haitian government presented a fully documented case with more than 200 witnesses appearing for the prosecution. In the opinion of the American minister, the bank's complicity was established beyond all reasonable doubt, though he felt the government erred by bringing its officers rather than the bank itself to trial. On Christmas Day, 1904, the jury delivered its verdict.

Found guilty and sentenced to various fines and up to four years in prison were the bank's director and several lesser officials. Involved members of ex-President Sam's administration were dealt with more severely. Sentences of up to life imprisonment at hard labor were meted out to ex-President Sam, ex-Minister of the Interior Tancrede Auguste, ex-Minister of War Vilbrun Guillaume Sam and ex-Minister of Public Works Cincinnatus Leconte.

A striking commentary on the times is that, of the guilty Haitian officials named, only one served any part of his sentence, and the last three named were later "elected" to the office of President of the Republic.

In July 1905, Minister Powell was recalled to Washington for consultation. There his involvement with "a syndicate of capitalists" was disclosed that cost him his office.

A disastrous defeat suffered by the Russian navy in the battle of Tsushima

5 Powell to Dept. of State, No. 1709, April 5, 1905.

Straits, May 27-28, 1905, left the Czar in need of ships to rebuild his fleet. It was patently impossible to build new ones in time to continue his war at sea against Japan, but one might be bought if the price was right. The syndicate Powell joined was interested and found two states willing to part with their navies under certain conditions. One was that the sale had to be indirect and completed through a third party in deference to their respect for international law and concern over possible Japanese reprisal, which was remote.

Thus it was that Powell was approached in Washington by a representative of the syndicate, which by chance was even then considering, at Powell's urging, a new consolidation of the Haitian debt. If the Haitian government would agree to buy the naval vessels of Chile and Argentina (with money advanced by the syndicate), and subsequently agree to transfer these ships to Russia at a place to be determined on the high seas, the syndicate would pay Haiti the sum of "$200,000 or $250,000" as soon as the sale was confirmed.

It seems probable that the vagueness of the syndicate's offer was deliberate, and the uncertain $50,000 reserved as a personal inducement to J.N. Legér, Haiti's minister in Washington, should that prove necessary. The receptiveness of Minister Legér to the proposal, made by Powell at a Sunday meeting, is not known, but he did agree to meet with the syndicate's representative the following Tuesday.

Powell began to have second thoughts, however, about how to best approach the Haitian government on this delicate matter. He wrote privately to the deputy consul in Port-au-Prince, a Mr. Battiste, and asked him unofficially to inform the minister of foreign relations of the substance of the syndicate's proposal and to state that the matter was then in the hands of Minister Legér.

Legér, when he found out what Powell had done, considered it, perhaps not without reason, an attempt to put himself in a compromising position with his government. He quickly, with high moral indignation, registered a public complaint with the U.S. State Department. He charged Powell with attempted bribery. What President Theodore Roosevelt, grooming himself to act as peacemaker in the Russo-Japanese War, thought about this airing of Powell's interest in sea power may well be imagined.

Powell was allowed to return to Haiti to clean out his desk, and his long years of service there ended on Nov. 25, 1905, more than a month prior to the arrival of his successor. Ironically, the Haitian navy was then awaiting delivery of the ex-*Prins Gegarin*, a 600-ton vessel that for 16 years had plied the Black Sea under the Russian flag. It was said to have cost a German agent 120,000 francs, and was sold to Haiti for $300,000 in American gold.

It is doubtful if any U.S. diplomat had ever stepped into a situation more fraught with chaos than did Henry W. Furniss, the new American minister in Port-au-Prince. A black man from Indiana of private means and shrewd intelligence, Furniss owed his appointment to the strident lobbying of African-American groups in the United States. Thus it was that the two most important State Department posts in Haiti were filled by black Americans throughout most of the decade preceding the U.S. intervention, for the consul at Cap Haitian, Lemuel W. Livingston, was also black. Both men were discerning and articulate diplomats, and their role in shaping American policy toward Haiti in this period can hardly be discounted.

Furniss established himself in a new and comfortable house in the suburbs of Port-au-Prince. He became an interested spectator of Haitian politics that flashed, kaleidoscope-like, through a bewildering procession of plots and counterplots. He bought a horse and traveled throughout the countryside, mastering the local patois. He studied the life and customs of the people. He found a mostly illiterate and apolitical peasantry ruled by venal and unscrupulous "generals" and politicians whose only goal seemed to be access to the treasury. This view, shared by Livingston, led both men to favor and encourage direct American action to right the crying wrongs of the Black Republic.

What Furniss reported concerning the evils of the Haitian political system and the sham of its "democratic" institutions, would be verified and reinforced by the statements and writings of dozens of equally qualified reporters in the years that lay ahead. It seems doubtful, though, that such reports ever significantly tarnished the popular myth of a small Caribbean republic fighting for liberty, freedom and justice in the clutches of foreign exploiters and would-be conquerors.

As has been shown, foreign exploiters there were, as well as those who coveted ultimate control of the land and the people, but Haiti's real distress always could be traced to failure of her indigenous ruling classes. That regrettable fact, however, did not comport with the reformers' and the zealots' images of Haiti, and it was predictable that reaction to the American intervention, when it came, would lead to demands for a return to democratic rights Haiti had never enjoyed.

The closing years of the Nord Alexis regime were distinguished by both the brutal and the bizarre. The president, then approaching the age of 90, seemed mentally imbalanced, and his derangement took on certain vicious characteristics that boded ill for those who opposed him or who aroused his suspicion. In September 1907 a military tribunal was established to try 17 people accused of conspiracy against the government. In his address to the court the prosecuting attorney said, among other things, that, "It was necessary that all the prisoners should be found guilty and shot, even if the government had to wade through blood to do so, in order that it might be shown to the government of the United States of America that they (the Haitians) were able to maintain order and were able to promptly put down all attempts at revolution without the intervention of that government."

The attorney for the defense, when he got an opportunity, said that the very illegality and arbitrary action of the court would, for the sake of humanity, appear to be reason enough for intervention by the United States. He was not permitted to go on with his address. The presiding officer of the court tapped a bell that caused bugles to blow and drums to beat, effectively drowning out whatever further was said.[6]

In less than half an hour, all the prisoners save one were found guilty and ordered shot. The "trial" illustrates both the quality of Haitian justice and what was plainly a growing preoccupation with the threat of U.S. intervention. The latter was, of course, quite understandable in view of American actions elsewhere in the Caribbean, most notably next door in the Dominican Republic. The American "menace" was a frequent topic of discussion.[7]

Political executions became so frequent that victims were jokingly said to

6 Furniss to Dept. of State, No. 244, Sept. 26, 1907.

7 Furniss to Dept. of State, No. 461, Dec. 4, 1908.

have contracted a "sudden disease," and this became an accepted euphemism for the murder for those the government wanted quietly out of the way. Foreign legations and consulates were filled with those seeking asylum, and at this moment the United States reversed its former liberal policy concerning the granting of asylum. Apparently convinced that abuse of the right to asylum was a contributing factor to political turmoil — and acting against the advice of Furniss, who had the backing of the entire diplomatic corps in Port-au-Prince — U.S. Secretary of State Elihu Root, in January 1908, ordered an end to the sheltering of Haitians in the American Legation and in all U.S. consulates in Haiti. When he later learned that local revolutionary leaders remained in the consulate at St. Marc, he summarily revoked that consul's commission.

The affair had an unhappy denouement, for no sooner was the American flag lowered that Haitian soldiers rushed in to drag out the refugees, who were marched to a local cemetery where, after digging their own graves, they were shot.[8] A helpless spectator to this was the commanding officer of the American gunboat *Eagle*, himself under verbal fire for earlier violating the new U.S. neutrality by preventing the Haitian navy from bombarding St. Marc.

Somewhat taken aback by the summary slaughter of the refugees, Root cabled Furniss instructions to hold any remaining Haitians seeking sanctuary "until absolute guarantees from the President and Government of Haiti can be obtained for their trial according to the orderly operation of Haitian law." This showed an appalling misconception of the true state of affairs in Haiti. Furniss courageously tried again to put things in proper perspective.

"Conditions in Haiti are such that it would be impossible to have any guarantees carried out as to legal and proper trials.... To consent ... to the delivery of the refugees ... to the Haitian government for trial would just be going through a formality that would in reality guarantee nothing for the accused. But even granting that the parties would have a fair trial and would be legally condemned, there is nothing to prevent their being cast into prison to be tortured, exposed and either made away with or be allowed to die of some lingering disease easily contracted in the prison. Under such conditions ... it would really be more

8 Commanding Officer USS *Eagle* letter 25-EJF, Feb. 5, 1908.

humane to deliver the refugees to the Haitian government and allow them to be as summarily dealt with as were those who took refuge in our consular agency at St. Marc ... My opinion as expressed is also the opinion of all my colleagues of the Diplomatic Corps." [9]

Despite its apparent cruelty and its faulty execution, the new American policy on asylum was unquestionably wise and long overdue. Consular asylum, though practiced for many years in volatile Latin American countries, had no generally accepted basis in international law. The existence of so many safe havens scattered throughout Haiti took much of the risk out of revolutionary activity. Among those expressing sorrow to Furniss over the new policy were distinguished members of the Nord Alexis administration, perhaps fearing the day when they themselves would have to flee for their lives.

By April 1908 the U.S. Legation and consulates in Haiti were entirely cleared of refugees, and most managed to find shelter in the legations and consulates of other foreign powers. Asylum ceased to be a matter of pressing concern to Minister Furniss. He had, however, other problems.

In April 1907 James Gordon Bennett, publisher of *The New York Herald,* visited Haiti with a group of friends while on a yacht cruise. By chance, he met a Haitian senator who took him to call on the president. He was pleased with his reception, and Haitian officials were particularly gratified with his treatment of them. Later, the government sent presents of coffee, fruits and other products of Haiti to Bennett, and the upshot was that a *Herald* correspondent was sent to "interview President Alexis and take sides with his Government."[10] The correspondent, a Mr. de Armas, spoke French fluently and his long service as a newspaperman made him naturally inquisitive. His investigation of Haitian affairs soon led him to the conclusion that it was impossible to defend the government. He cabled his observations to "Bennett, who instructed him to tell the truth but to be as easy as possible on the present government."[11]

The *Herald* stories, at first read with eager pleasure by government officials

9 Furniss to Dept. of State, No. 299, Feb. 4, 1908.

10 Furniss to Dept. of State, No. 431, Sept. 10, 1908.

11 Ibid.

in Port-au-Prince, soon soured. Mr. de Armas left Haiti in fear for his life. His place was taken over by an undercover correspondent. Sensationalism crept into the accounts that began to appear in the pages of the newspaper. Speculation about possible American intervention, and editorial comment that seemed to urge that course of action, enraged the Nord Alexis regime while strengthening the opposition, whose pledge it now was to save Haiti from the "American peril."

How much the *Herald* stories sped the downhill slide of the Haitian government is a matter of conjecture. By November 1908 a full-scale revolt was under way, and its leader was Antoine Simon, for 18 years governor general of the Department of the South. Government troops sent forth to do battle with Simon's forces were literally tied together to prevent their desertion. Furniss and his French and German colleagues urged joint intervention to protect foreign interests and put an end to disorder. U.S. firms doing business in Haiti bombarded the State Department with pleas to send troops. From Secretary of State Root came a terse and unequivocal reply: "The United States has no idea of intervening in Haiti."[12]

The day before Root's cable to Furniss, President Theodore Roosevelt sent a personal letter to William Bayard Hale, editor of the *Saturday Review of Books*. Roosevelt gave his take on the situation in Haiti:

"The trouble as regards Haiti is not that the Government fails to understand the situation but that the people, especially the educated people, refuse to understand it. The *Times and Evening Post,* for instance, which was lukewarm or hostile to what I did about Santo Domingo — and which nothing so good for Santo Domingo has ever been done — I got it thru the Senate by two years hard work, without a particle of aid from what likes to call itself the 'educated conscience of the people' of the country. Now, in Haiti, what we need is something that will show our people that this Government, in the name of humanity, morality and civilization, ought to exercise some kind of supervision over the island; but this should be done as a part of our general scheme of dealing with the countries around the Caribbean. In Cuba, Santo

12 Secretary of State to Furniss, Cable, Dec. 4, 1908.

Domingo and Panama we have interfered in various different ways, and in each case for the immeasurable betterment of the people. I would have interfered in some similar fashion in Venezuela, in at least one Central American state and in Haiti already, simply in the interest of civilization, if I could have waked up our people so that they would back a reasonable and intelligent foreign policy which would put a stop to the crying disorder at our very doors. Such a policy would be a little in our own interests, but much more in the interest of the peoples in whose affairs we interfered. I think Mexico would have gone in with us on a rational policy. But in each case where I have actually interfered — Cuba, Santo Domingo and Panama, for instance — I have had to exercise the greatest care in order to keep public opinion here with me so as to make my interference effective, and I have been able to lead it along as it ought to be led by minimizing my interference and showing the clearest necessity for it. In the other cases I have mentioned, tho the need was great, it was not as great as those in which I did take action, and the need could not have been as clearly shown to our people. Our prime necessity is that public opinion should be properly educated...."[13]

President Nord Alexis, 91 and terribly frail, left the palace at Port-au-Prince at five in the morning on Sept. 2, 1908. Accompanied by the French minister and a cavalry escort, he proceeded to the harbor where he embarked on a French warship. Despite the early hour, large crowds had gathered where it was supposed he would pass. Several persons tried to strike him, and shots were fired at his entourage. The crowd took possession of his baggage, breaking open the trunks and completely emptying them out, taking all of his clothing, papers and money.

After he left, there was considerable "jollification" in the city. Many who had guns spent the greater part of the day and night shooting into the air. The whistle of bullets near the American Legation became so common that Minister Furniss considered it unsafe even to sit on the back porch of the building. He may be forgiven if he viewed the fall of the government with an overall feeling of relief.

13 Theodore Roosevelt to Wm. Bayard Hale, letter dated Dec. 3, 1908.

On the morrow, however, Simon's army clattered through the streets of Port-au-Prince and ushered in a new and almost incredible era in Haitian history.

6

THE ERA OF THE EPHEMERAL GOVERNMENTS

As Bobo and his four frock-coated dusky followers mounted the ladder, it was observed that all carried their suitcases upon which had been painted: Doctor Rosalvo Bobo, Chief of the Executive Power," "Doctor So and So, Minister of the Interior," etc.

— Commander C.S. Baker, Supply Corps, U.S. Navy.

The suitcases observed in the hands of an aspiring Haitian administration as it boarded an American ship in the harbor of Port-au-Prince in 1915 were appropriate symbols of the political history of Haiti in the last years prior to the American intervention. On Dec. 17, 1908, Antoine Simon was elected to the presidency for the constitutional term of seven years. In the seven years following that election no fewer than seven presidents were elected and deposed in Haiti. Simon himself served the longest, two years and seven months. The rapidity of the turnover thereafter well explains why the Haitians call this period the "*Epoque des Governements Ephémères*."

President Simon added flair, an artistry, to the plunder of the state treasury. He had a large family to provide for, and all his sons were given important offices where the opportunity for pilfering was great. His daughter, Celestine, became a grey eminence in the palace, openly practiced voodoo and peddled influence. The president's "perquisites" — pay and allowances for a regiment that didn't exist, "secret funds," "rations" etc. — amounted to a little more than $1 million a year in addition to his official salary of $25,000. On his birthday, $25,000 was appropriated for fireworks used in the celebration, and a skepti-

Haitian President Antoine Simon.

cal Furniss reported that $100 would have more than covered the expenses.

In September 1910 a special law was passed setting aside $50,000 in gold for the purchase of a plantation for the president "in return for amounts which he has to personally expend on various occasions for the public good."[1]

To the Simon administration fell the task of reorganizing the National Bank, then increasingly in difficulty due to rampant corruption in the customs service. In February 1909 the bank proposed a new contract that would give to it responsibility for collecting revenues and apportioning disbursements. The U.S. State Department characterized it as "a business proposition under which officials of the bank may be enabled to assume control over the political affairs of the State, a condition which would tend to result in the complete subjugation of Haitian political rights to French commercial interests and the increased influence of a private corporation at the expense of the nation."[2]

Germany, through its Dresden Bank, tried to buy out French interests. Failing that, it urged Simon to force the National Bank into receivership. With the encouragement of the U.S. State Department, a group of New York firms made a bid.

The eventual reorganization of the National Bank of Haiti in August 1910 saw 74 percent of its stock going to its original French holders, 6 percent to the Dresden Bank and 5 percent each to the American firms of National City Bank, Messrs. Speyer & Co., Hallgarten & Co., and Ladenburg, Thalman & Co., the last three being houses of close German affiliation. Along with the reorganization went a 5 percent loan of $16 million, of which the Haitian government netted only 72.3 percent. The low rate of issue is explained not only because of Haiti's low credit rating, but personal inducements of at least $5 million paid to the Haitian officials who signed the agreement.

Money realized from the loan was to be used to settle accounts with the old bank, refund the internal debt and retire the fiduciary currency, for which purpose some $2.5 million was reserved. The loan was secured by a tax of $1 in gold on each 100 pounds of coffee exported, and a special 15 percent surtax on imports. Collection was left in the hands of the government, although the bank controlled service of the public debt, which was the sole depository of the Haitian government.

1 U.S. National Archives, Furniss to Dept. of State, No. 731, Sept. 17, 1910.

2 Dept. of State to Furniss, Feb. 23, 1909.

In 1910 the National Railroad of Haiti was reorganized under the directorship of colorful financier and adventurer James P. MacDonald, who allegedly secured valuable concessions by charming President Simon's daughter, Celestine. Under the terms of a new contract, the Haitian government guaranteed 6 percent interest on railroad construction bonds up to a value of $35,000 per mile of finished track. New York's National City Bank loaned MacDonald $500,000 to start construction. The railroad was to run from Port-au-Prince to Cap Haitian, and National City Bank received as collateral for its loan the bonds of the railroad.

National City Bank already had an interest in Haiti's other railroad company, the Central Railway of Haiti. It controlled the Plaine du Cul de Sac (PCS) Railroad, which ran from Port-au-Prince to Lake Sumatre, a distance of about 28 miles.

The National Railroad concession set a pattern for other foreign investments in Haiti, and government default on guaranteed interest payments would lead to heavy pressure for diplomatic and, eventually, military intervention.

On Feb. 25, 1911 the American minister in Port-au-Prince cabled Washington that, "The Simon Government, by its summary executions, merciless slaughters, burlesque of liberty, misappropriations of government funds and other unconstitutional acts, can now be regarded as vile a government as has ever existed in Haiti."[3]

That summer, revolutionary activity led by Cincinnatus Leconte broke out in the interior of the country. For a time the situation was, in a word, confused. In July the commanding general of forces in Valliere was ordered to the front. During the night his forces were attacked and he retreated (or more correctly ran) with his soldiers back to Valliere. The general commanding forces from Hinche, seeing the fleeing troops, mistook them for the advancing enemy. He and his troops fled to Gonaives, where he reported that Hinche had fallen. When the truth became known, the whole affair was thought to be a great joke.[4]

Leconte's real advance came later in July and President Simon, convinced

3 Furniss to Dept. of State, No. 835, Feb. 25, 1911.

4 Furniss to Dept. of State, No. 898, July 18, 1911.

that the country no longer was with him, embarked with his family on Aug. 2, 1911. His departure was made under cover of the German-flagged cruiser *Bremen,* which landed 40 men. Celestine was shot and slightly wounded on her way to the harbor.

The senior American naval officer present at Port-au-Prince received instructions to land forces as necessary to prevent looting and to protect U.S. lives and property. This proved to be unnecessary, probably because of a joint warning by the diplomatic corps to the generals commanding revolutionary forces that fighting and disorder would not be permitted in Port-au-Prince. With numerous foreign warships then in the harbor, the warning was heeded and Leconte was proclaimed "Chief of the Executive Power" in a relatively quiet ceremony. On Aug. 4 he was elected Haiti's president, unanimously. The German business community subscribed an immediate loan of $200,000 to the government. Leconte's subsequent pro-German leanings were viewed with distrust by other foreign interests in Haiti.

Almost exactly one year later, on Aug. 8, 1912, President Leconte died in an explosion of a powder magazine on the palace grounds. The magazine was a one-story building with a thin, brick wall adjoining the palace and a barracks for the palace guard. The explosion was attributed to a failure to ventilate a storage chamber for black powder. The disaster occurred at 3 a.m. Leconte was trapped in his bed by debris and burned to death. His son, the minister of public works, also perished along with 100 palace guards.

The so-called Syrians, those who remained from the Nord Alexis purges of several years previously, were made scapegoats, and there were scattered cases of looting and arson in the city.

Leconte was succeeded by another figure in the consolidation scandal, Gen. Tancrede Auguste. His election occurred with almost indecent haste on the day of Leconte's death. President Auguste's rule, which lasted just nine months, is most remarkable for the events surrounding his funeral and the choice of his successor. It was popularly thought his death was caused by slow poisoning, and that the party responsible was his mistress.

Minister Furniss, who was a collector of backstairs gossip, reported that the president died in the terminal stages of syphilis (in which case, of course, both versions of his death may be correct!)[5]

The funeral service began at 8:30 a.m. on Sunday, May 4, 1913. It was held at the Port-au-Prince Cathedral. Minister Furniss arrived on time and found his German, French and British colleagues, in full diplomatic regalia, seated immediately in front of pews reserved for the Haitian Cabinet. Senators and deputies were seated across the aisle facing the diplomats. The cathedral soon filled and after a short time the funeral cortege arrived. The president's body was placed upon a bier, and the Cabinet, surrounded by officers armed with 16-shot Winchesters, took their assigned places.

As was customary at *Te Deum* masses in which Haitian officials took part, armed troops were drawn up in the square facing the cathedral. The mass progressed with the Archbishop delivering a sermon in which he urged that there be no strife in the choosing of a new president, reminding legislators of their duty to the nation and expressing hope that progress started by Auguste would continue.

Clergy and church assistants then formed a procession. At the funeral bier, just inside the front door of the cathedral, they began to give the last rites. At that moment the crack of a rifle resounded in the square, quickly followed by a volley from all sides.

The whole congregation was thrown into panic. Those near the doors attempted to gain the center of the cathedral, while those near the main aisle stampeded in an attempt to reach anterooms behind the altar. Furniss' German, French and British colleagues mounted the dais in front of the altar, but Furniss, aware that Haitians more often than not fired into the air rather than at anyone, wisely took shelter in another part of the cathedral, away from the vicinity of the Cabinet officers. Soldiers fired into the cathedral's ceiling while shouting *"A bas le Cabinet!"*

The initial volleys did not last long, but continued intermittent shots seemed to threaten fresh outbreaks, and it was some time before many dared to venture out. Friends advised Furniss that there was a growing party in Haiti that believed American intervention offered the only salvation for the country, and

5 Furniss to Dept. of State, No. 1217, April 29, 1913.

that an attack on the American minister might bring it about. So he delayed his departure from the site until he was convinced that the neighborhood near the cathedral was quite deserted. Edging his way to a side door, he then slipped out and ran the three blocks to the U.S. Legation.[6]

Tancrede Auguste's death resulted in the "election" of the first civilian president in Haiti's history. Sen. Michel Oreste, a prominent lawyer, was chosen by a vote of 72 to 24 shortly after the funeral ceremonies. His candidacy was unquestionably advanced by an open distribution of promissory notes to the electors of from $300 to $1,000 payable after the election. One of the first acts of the new administration was the introduction of a law appropriating $126,000 in gold for "extraordinary expenses as a result of the death of President Tancrede Auguste." The law passed handily.[7]

In the summer of 1913, coincident to the recent change of administration in Washington, Furniss was replaced as minister in Haiti by Madison R. Smith. This change was unfortunate, since Smith proved to be unsuitable for the assignment. During his tenure, negotiations were increasingly placed in the hands of special emissaries from the Wilson administration. Furniss stayed on in Port-au-Prince as a private citizen.

In July 1913, Gov. John Eugene Osbourne, recent appointee as assistant U.S. secretary of state, secured an agreement from the Oreste government that Mole St. Nicholas would not be sold or leased to any other foreign country. Whether such agreement had any real worth, given the mercurial condition of Haitian politics, is doubtful. A new revolution, nominally led by Davilmar Theodore, was already under way.

Extraordinary measures were taken to shore up the Oreste regime, including the use of USS *Nashville* to carry government forces to Port-au-Prince in the face of rebellion. U.S. Secretary of State William Jennings Bryan cabled Minister Smith that, "the Department is disposed to do anything proper to support constituted government."[8] The revolt against Oreste carried special

6 Furniss to Dept. of State, No. 1223, May 10, 1913.

7 Furniss to Dept. of State, No. 1236, May 28, 1913.

8 Secretary of State to Smith, Cable, Jan. 22, 1914.

USS *Montana.*

significance since it was the first such incident since the announcement of President Wilson's Latin American Policy in Mobile, Ala. on Oct. 27, 1913.

In spite of American support, Oreste abdicated on Jan. 27, 1914, and embarked on a German warship. German and American landing parties kept disorder in Port-au-Prince in check. The USS *Montana* sent some 150 men ashore.

The Theodore revolt was derailed en-route to the capital by the Zamor brothers — Oreste and Charles. On Feb. 7, 1914, they entered Port-au-Prince at the head of an army of 4,000. Oreste Zamor was elected president the next day. Charles Zamor was named minister of the interior. Theodore, meanwhile, unfurled the banner of revolution once more by proclaiming himself "Chief of the Executive Power" from his stronghold at Cap Haitian.

When the Zamor brothers came into power they found, not surprisingly, an empty treasury. Further, they were without the means to effect the necessary reforms in the customs service that would assure them a future source of revenue. The National Bank refused to advance them any funds because the government was unable to provide security or collateral of any kind.

This enraged President Zamor and threats were made against the bank,

U.S. Secretary of State William Jennings Bryan.

threats that were answered by the French minister, who officially notified the Haitian minister of finance that, if the threats against the bank were executed, "the French Government would take very drastic action without regard to the attitude of the United States."[9] This was quickly followed by demands

9 U.S. Dept. of State, Internal Memorandum of the Latin-American Division, March 25, 1914.

from both France and Germany for participation in the administration of the customs houses in the country.

At this moment a series of interesting exchanges occurred between Secretary of State Bryan and R.L. Farnam, an employee and later vice president of the National City Bank of New York. Farnam also served at this time as president of the National Railroad Company of Haiti, and thus had numerous connections and sources of information in both the railroad and the National Bank of Haiti. Bryan, who seemed to harbor latent distrust of officials in his own department, was on the receiving end of advice passed to him by Farnam in the early months of 1914. It reached him by letter, by telegram, by telephone and in several face-to-face meetings. Farnam frequently prefaced his communications to the secretary of state with an expression of hope that he would present a view of the situation in Haiti "somewhat different from the view one obtains from the regular official reports."[10]

Farnam described President Zamor as savage, inebriate and threatening to the National Bank and Railroad. The only solution to the Haitian problem, Farnhan said, was the separation of the customs service from politics. He further purposed to reorganize the National Bank as a corporation that would be 40 percent American, 50 percent French and 10 percent German, with the new bank entrusted with collection of the customs and discharge of the public debt. An essential element in Farnham's plan was guaranteed protection of the bank in the execution of its expanded duties by the United States.

The Latin American affairs division of the U.S. Department of State was less than enthusiastic in support of Farnham's proposal, not only because of the minority position of American interest in the prospective corporation but because an advance guarantee represented a significant departure from the prior policy of deciding such matters on a case-by-case basis and on the merits. Nevertheless, Farnham, acting as the personal representative of Secretary of State Bryan, went to Haiti in July 1914. His mission was to secure the consent of the Zamor brothers to his proposal.

Meanwhile, revolutionary activity in the north of Haiti, aided by clandes-

10 Personal letter from R. L. Farnam to William Jennings Bryan, March 11, 1914.

tine shipments of arms and munitions across the border from the Dominican Republic, increased drastically. An interesting sidelight to this, and perhaps an example of "CIA" activity long before there was a CIA, led to the Zamors' purchase of a 3-inch field piece in New York and the hiring of a former U.S. Navy lieutenant and a former Navy gunner's mate to assist in the reduction of a rebel fortification at the northern town of Ouanaminthe. The field piece and the former naval personnel rode to Haiti on the same ship with Ron Hazeltine, a State Department consular assistant on special detail.

The war in Europe made improbable any reorganization of Haiti's National Bank such as that envisioned by Farnham. After several weeks in which he commandeered U.S. warships, gave instructions to State Department representatives there and generally made himself a nuisance, he returned to the United States.

In the interim, Smith was relieved as minister by the equally unenergetic Arthur Bailly-Blanchard, who arrived in Haiti with an alternative proposal that provided for U.S. supervision of the customs by an American-nominated and Haitian-appointed financial advisor. The plan was embodied in the draft of a proposed convention that followed closely one already in effect with the Dominican Republic. Not coincidentally, the USS *Hancock* and 800 Marines were ordered to Guantanamo in readiness for duty in Haiti.

The Zamors agreed to the convention as a last desperate measure to remain in power. They insisted, however, that the United States land its forces in Port-au-Prince before treaty negotiations began in order to take the edge off opposition charges that they were preparing to sell their country to the Americans. Secretary Bryan remained firm in his conviction that the treaty would be adopted only on the prior request of the Haitian government for friendly American assistance.

The internal situation in Haiti steadily worsened. Political prisoners were paraded through the streets of Cap Haitian and shot amid cries of "*Viva Zamor!*" Opposition leaders told Consul Livingston they would rather see themselves and their country perish 10 times over than be governed by the Zamor administration.

In August the Haitian government abrogated its agreement with the National Bank and sued the National Railroad, steps that brought cries of outrage

Haitian President Oreste Zamor.

from Farnham and stern warnings from Secretary Bryan. In September soldiers in Port-au-Prince threatened the Zamors with death unless salaries were paid. They were then given permission to take food from the markets. The country tottered on the brink of anarchy.

On Oct. 19, 1914, the USS *Tacoma*, at the urging of foreign consuls in Cap Haitian, sent a landing party ashore to protect foreign interests in face of the imminent fall of the city to revolutionists. A few days later, President Zamor agreed to have the views of the U.S. government adopted in exchange for U.S. assistance in fighting the revolution.[11]

Acting Secretary of State Robert Lansing sent an urgent request to the White House for instructions concerning Haiti. Shortly thereafter the following cable was sent to Port-au-Prince:

"Transport *Hancock* with 800 Marines has been ordered to proceed to Port-au-Prince immediately to land men at your request. *Hancock* is due at Port-au-Prince the evening of Oct. 30. Battleship *Kansas* from Mexican waters has also been ordered to Port-au-Prince.

"Department is still contemplating convention along lines of its instructions of July 2. At the same time it is desired to reach agreement for a fair election to be held within a specified time similar with the Department's action in the Dominican Republic. Department feels it advisable to negotiate with the Zamor brothers, and if President Zamor has already left the country, wishes every protection offered Charles Zamor now reported refugee in Consulate.

11 Baily-Blanchard to Dept. of State, Cable, Oct. 25, 1914.

Upon arrival of *Hancock* you will at your discretion request Commanding Officer to take charge of Port-au-Prince and will restore Charles Zamor to his cabinet functions. Further instructions will be sent shortly. Report by cable."[12]

Thus the decision to intervene had been taken, at the belated "request" of Charles Zamor. The American response, however, was not received in time, for Lansing's cable crossed another one from Bailly-Blanchard that reported a general uprising in Port-au-Prince during which the Zamors had fled, the president to a Dutch vessel and Charles to diplomatic asylum in the French Legation. When the *Hancock* and the Marines arrived on the morning of Oct. 31, calm reigned in the city. No occasion could be found to land troops.

On Nov. 6, Davilmar Theodore arrived in the capital amid the ringing of bells and the firing of a presidential salute. The next day he was elected "constitutionally" President of the Republic.

Secretary of State Bryan cabled instructions for the American minister to advise Theodore that the United States would recognize him as a "provisional" president when he named a three-man commission to go to Washington to negotiate a convention or conventions covering: (1) U.S. control of the Haitian customs service, (2) settlement of all matters outstanding between the government of Haiti and the railroad, (3) settlement of questions outstanding between the government of Haiti and the National Bank, (4) agreement of Haiti to give full protection to all foreign interests there, (5) agreement by Haiti never to lease any territory at Mole St. Nicholas or anywhere else in the country to any European government for use as a naval coaling station and (6) appointment of a Haitian commission empowered to sign a protocol for use in settlement by arbitration of pending claims of American citizens against Haiti.[13]

The Theodore government resisted Bryan's demands, and made a counterproposal to grant Americans special privileges in mining and other operations. Bryan replied that what was required to attract American capital was peace and stability, not special privilege. Strongly pressed, an attempt was made to open the question of customs control, but on Dec. 3, the Haitian minister of

12 Acting Secretary of State to Bailly-Blanchard, Cable, Oct. 29, 1914.

13 Secretary of State to Bailly-Blanchard, Cable, Nov.. 12, 1914.

foreign affairs, while undergoing questioning on the subject, was physically attacked in the Haitian senate chambers and accused of trying to "sell" the nation to the United States. He departed amidst the wildest excitement and later submitted his resignation.

In this crisis, Farnham hurried to Washington, and in a meeting with Bryan secured his consent for an extraordinary transfer of $500,000 in gold — part of the funds reserved under terms of the National Bank's contract for the retirement of the old Haitian paper currency — from the vaults of the Haitian National Bank to the National City Bank in New York. Because this was expected to meet with strong resistance, arrangements were made to affect the transfer using an American naval vessel and American Marines.

The USS *Machias* was ordered to Port-au-Prince, and on Dec. 17, 1914, a party of eight Marines armed with stout canes and concealed revolvers transferred the gold from the National Bank to the ship without incident. The American officer in charge of the operation wisely chose the Haitian siesta hour for the transfer, and the gold passed through nearly deserted streets to the harbor. Both the U.S. minister and the local bank manager strongly protested Secretary Bryan's decision in this matter.

In mid-January 1915, having refused recognition of the Theodore government, President Wilson decided on a new initiative and wrote Secretary of State Bryan, as follows:

"... The more I think about that situation the more I am convinced that it is our duty to take immediate action there such as we took in Santo Domingo. I mean to send a commissioner there who will seek and obtain an interview with the leaders of the various contending factions of the Republic and say to them as firmly and definitely as is consistent with courtesy and kindness that the United States cannot consent to stand by and permit revolutionary conditions constantly to exist there. They might, as in Santo Domingo, insist upon an agreement for a popular election under our supervision and be told that the result of that election would be upheld by the United States to the utmost.

"Is this not your judgement?"[14]

14 Woodrow Wilson to Secretary of State, Memorandum, Jan. 13, 1915.

Rear Adm. William Banks Caperton.

Apparently that was Bryan's judgment. He recommended that Gov. John Franklin Fort, Charles Cogswell Smith (both of whom had recently acted as commissioners to Santo Domingo) and Arthur Bailly-Blanchard comprise the U.S. Haitian Commission. Wilson agreed.

The situation in Haiti was precarious and it looked as though the Haitian government might fall at any moment. The leader of the new revolution was Vilbrun Guillaume Sam, and his activities in the vicinity of Cap Haitian caused Consul Livingston to request the presence of a U.S. Navy vessel. Rear Adm. William B. Caperton, commander of Cruiser Squadron, Atlantic Fleet, arrived there with USS *Washington* on Jan. 23, 1915. Shortly thereafter the admiral and Livingston called upon Sam. The meeting was a memorable one.

Years later Caperton recalled that a "very gorgeous black gentleman arrayed like a head bellhop at the Waldorf" greeted them at the door to Sam's headquarters and directed them across a room and to a short flight of stairs. The gentleman then disappeared only to appear again, still resplendent, at the head of the stairs where he took their caps and gloves and ushered them into a large reception room, saying the general would be pleased to see them in a moment. He then disappeared again. While the admiral waited, a large portrait caught his eye. It seemed familiar. When Gen. Sam appeared, the admiral was surprised to recognize him as the "bellhop," only this time he had discarded his coat for a more elaborate one, and an enormous sword clanked round his heels. A glance at the portrait on the wall convinced Caperton that it also was Guillaume Sam.[15]

Livingston had described to the admiral the usual course of Haitian revolutions. During Adm. Caperton's interview with Gen. Sam he extracted a promise that as Sam advanced with his army to Port-au-Prince there would be no looting of the towns along the way. He arranged to have naval officers meet Sam on each major stop on his way to the capital to remind him of his promise. The admiral then departed Cap Haitian on the *Washington* and proceeded to Port-au-Prince, arriving on Jan. 27. He discovered there a tense situation involving the American schooner *Alice Pendleton* and the Haitian government.

President Theodore had proclaimed a blockade of the ports of northern Haiti, but was unable to enforce it because Haitian naval vessels were without coal, and funds to purchase any. The *Alice Pendleton* had been at Port-au-Prince since Jan. 11 with 600 tons of coal contracted but not yet paid for by the Haitian government. When Caperton arrived he found the schooner alongside a wharf between the Haitian gunboats *Nord Alexis* and *Pacifique.*

In view of the financial condition of the Theodore government — its urgent and vital need for funds, the lack of coal for its gunboats, the exposed position of the American schooner and the approach of the revolutionists — Caperton ordered the *Washington* to ready its landing party. He also kept in constant touch with U.S. Minister Bailly-Blanchard.

On Jan. 29 the Theodore government published a proclamation changing the depository for customs receipts from the National Bank to such business houses or banks as it might designate. That same day, payment having been advanced, the *Alice Pendleton* began transferring coal to the Haitian gunboats. Renewed threats were made to forcibly remove from the National Bank funds already on deposit there. These were countered by a strong warning issued by the American minister, a warning underscored by the arrival on Jan. 31 of the USS *Montana* with 650 U.S. Marines. The decision was made, however, to delay any landing until after the arrival of the U.S. Haitian Commission.

On Feb. 22 the Theodore government fell. The Haitian president embarked

15 Adm. William B. Caperton, "Activities While in Command of Cruiser Squadron, U.S. Atlantic Fleet and Commander in Chief, U.S. Pacific Fleet." Unpublished manuscript, the Library of Congress.

on a Dutch ship, and Gen. Sam arrived in Port-au-Prince to the ritualistic ringing of bells and the cheering of the populace. This was followed by an equally ritualistic election on March 4. Gov. Fort and Smith landed the next day. Minister Bailly-Blanchard, of course, was already there.

President Wilson's Haitian Commission accomplished nothing during its 10-day sojourn in Port-au-Prince. Fort found his accommodations unsanitary and unsafe, and complained bitterly about the lack of a U.S. warship in the harbor during part of the commission's stay. Brief meetings were held with President Sam and his minister of foreign affairs. These ended on a sour note when the Haitian minister asked to see the commission's credentials and they could not be produced, U.S. Secretary of State Bryan having neglected to provide any.

After a week in the country, Gov. Fort reported that the commission was in full possession of the facts and that a longer stay would not be productive. He and Smith embarked on USS *Nashville* from Port-au-Prince on March 15. Bailly-Blanchard did not accompany them, and apparently did not anticipate Bryan's cable after Fort had reported to President Wilson. The cable stated in part:

"The President regrets to learn that the American Minister to Haiti has disregarded the instructions given to him by this Government and has not only failed to perform his duty as a member of the Commission but has on the contrary, used his influence to obstruct the work entrusted to the Commission. A full and immediate explanation of his attitude and his reasons therefore is requested."[16]

This severe reprimand apparently was more the result of Fort's ill-temper than anything done or left undone by Bailly-Blanchard, for after a hasty trip to Washington and explanations, the U.S. minister was returned to his post. His absence would cause him to miss, however, a moment of high drama in Haitian history.

In retrospect, the Fort mission was poorly timed and it is doubtful that even the most skilled negotiator, which Gov. Fort patently was not, could have achieved President Wilson's ends. Having just assumed office, President Sam

16 Secretary of State to Bailly-Blanchard.,Cable, March 26, 1915.

had a breathing spell, albeit a short one, before the next revolution began. His most pressing financial problems were eased by large cash loans from his uncle, former President Simon Sam, and by a lucky stroke. The man he considered his chief potential rival, Oreste Zamor, was arrested shortly after entering Haiti from the Dominican Republic and imprisoned to Port-au-Prince.

President Wilson was not of a mind to wait for the onset of a new crisis in Haiti. He chose a new and more capable emissary in Port-au-Prince, Paul Fuller Jr., a member of a prominent New York law firm. Carrying credentials as Presidential Envoy Extraordinary and Minister Plenipotentiary, Fuller arrived in Port-au-Prince in May. He immediately set about to negotiate a treaty skirting the sensitive issue of American financial control. He had detailed instructions from President Wilson himself. The most memorable and, in light of later U.S. experience, most poignant part of his instructions was the admonition that "only an honest and efficient government deserves support. The Government of the United States could not justify the expenditure of money or the sacrifice of American lives in support of any other kind of government."[17] Such a government, of course, did not exist nor had it ever existed in Haiti.

Fuller's draft convention provided for the posting of an American minister plenipotentiary in Haiti who would advise the President of Haiti "as to such matters as effect the honest and efficient administration of the Government, the President of Haiti agreeing that he will follow the advice so given to the extent of requiring honesty and efficiency of officials and of removing those found to be dishonest and inefficient."[18] President Wilson later directed that "dishonest and inefficient" be changed to read "dishonest or inefficient."

The question raised by the section quoted is, when does advice cease to be advice and become something more? It would seem that point is reached when it is agreed to follow, in advance, the "advice" given.

The draft convention contained four specific agreements in addition to that cited above. These were: (1) The Government of the United States of America will protect the Republic of Haiti from outside attack, and from the aggression

17 Woodrow Willson to Paul Fuller, Jr. May 6, 1915.

18 Paul Fuller Jr. to Sec. of State Cable, May 22, 1915.

of any foreign power, and to that end will employ such forces of the Army and Navy of the United States as may be necessary. (2) The Government of the United States of America will aid the government of the Republic of Haiti to suppress insurrection from within and will give effective support by the employment of the armed forces of the United States Army and Navy to the extent needed. (3) The President of the Republic of Haiti covenants that no rights, privileges or facilities of any description whatsoever will be granted, sold, leased or otherwise accorded directly or indirectly by the Government of Haiti concerning the occupation or use of Mole St. Nicholas to any other foreign Government or to a national or nationals of any other foreign Government. (4) The President of the Republic of Haiti covenants that within six months of the signing of this Convention, the Government of Haiti will enter into an arbitration agreement for the settlement of such claims as American citizens or other foreigners may have against the Government of Haiti, such arbitration agreement to provide for the equal treatment of all foreigners to the end that the people of Haiti may have the benefit of competition between the nationals of all countries. [19]

Two major issues confronted Fuller in his discussions with the Haitian government. The first of these concerned recognition, for though a special representative of the United States was formally engaged in treaty negotiations with the Sam government, that government had not been recognized by the United States. Sam insisted on recognition before signing the treaty, and President Wilson just as steadfastly insisted on a treaty before granting recognition. On May 28, Secretary of State Bryan cabled Fuller saying "the President authorizes recognition before signing providing you have his (Sam's) written promise to sign treaty immediately after recognition." [20]

The real stumbling block in the negotiations, however, concerned the terms under which American military forces would be permitted to operate in Haiti. It was the contention of the Sam government that American forces would have to withdraw at the first demand of constitutional authorities. To this

19 Paul Fuller Jr. to Sec. of State, cable May 22, 1915.

20 Sec. of State to Paul Fuller Jr. Cable, May 28, 1915.

Haitian President Vilbrun Guillaume Sam.

Fuller would not agree, stating that such a provision in effect would place the armed forces of the United States at the beck and call of a foreign government. Sam replied that while he himself did not doubt the good intentions of the United States, intervention was a very sensitive subject with the Haitian people. His counter-proposal on troop employment therefore was a final one. On June 15, Fuller left Haiti, the convention unsigned.

Meanwhile, revolution had broken out again in the north, this time under the eccentric leadership of Dr. Rosalvo Bobo, a former minister in the Theodore government. In June he took temporary control of Cap Haitian, looted the customs house and paraded the heads of several government officials through the streets. Threats were made to forcibly remove refugees from the French consulate, and 50 sailors were promptly landed from the French cruiser *Descartes.*

Convinced there was a widespread conspiracy against him, President Sam ordered sweeping arrests in Port-au-Prince. Some 200 of his opponents, many of them from upper class Haitian families, were thrown into the National Prison. Foreign legations and consulates, with the exception of the American, were again filled to overflowing with asylum seekers. The deteriorating political situation, and more particularly the landing of French sailors at the Cap, led to the ordering of Adm. Caperton, then in Mexican waters, to proceed at once to Haiti.

Thus it was that the USS *Washington*, after a midnight provisioning, got under way at the most unlikely hour of 2:25 a.m. on June 26 from Progreso, Mexico. Steaming on 16 boilers, she headed east across the Gulf of Mexico toward Haiti. After a brief fueling stop at Guantanamo Bay, Caperton and the *Washington* continued on to Cap Haitian, arriving there at 9:27 in the morning of July 1, 1915.

Within 10 minutes of anchoring, a salute of 13 guns was exchanged with the *Descartes,* which, as Adm. Caperton learned, had withdrawn its landing party a week earlier. The commanding officer of the *Descartes* placed his ship and crew at the admiral's disposal, but in keeping with instructions, Caperton simply thanked him for his services and informed him that henceforth he would assume responsibility for the safeguarding of foreign interests at the Cap.

Adm. Caperton, accompanied by Consul Livingston, next conferred with the French Consul and the local commander of Haitian government forces. He then proclaimed that he would not permit any fighting in the city, and that any battles between the two contending parties would have to be fought "well clear of town." This was necessary, he said, to carry out his orders to protect foreign lives and property, and to that end, should it become necessary, he was prepared to land American forces. He further declared that in the present political difficulties in Haiti, the United States intended to remain impartial. A copy of the admiral's proclamation was delivered to Dr. Bobo, who was encamped with his troops on the outskirts of the city.

On July 4, the *Washington* was joined at Cap Haitian by the gunboat USS *Eagle*, which arrived from Guantanamo by way of Port-au-Prince where, for the moment, conditions seemed stable.

On July 9 one officer and 29 Marines were landed by the *Washington* to quell a disturbance in the town. They remained there to set up a patrol of the streets. Small detachments were posted at the American consulate, the railroad station and the telegraph office. From then until July 27, the two American ships maintained a close watch on the threatening situation at Cap Haitian.

The *Eagle* took on coal and provisions, including: 48 garters, Peary lisle; 24 mouthpieces for clay pipes; 5,000 Melanchrino cigarettes; 432 Hershey Almond Bars; and 72 packages of playing cards — all carefully noted in the

USS *Washington*, Armored Cruiser No. 11.

deck log. A Marine private on the *Washington* was sentenced to a bad conduct discharge and a loss of pay in the amount of $68.80 for using excessively obscene language (brilliantly recorded for posterity) to the corporal of the guard, and on the same ship a lowly coal passer was awarded a deck court martial for "refusing to obey a lawful order, in that he did, when ordered to get a bucket of drinking water for the fire room watch, refused to do so."[21] Discipline in the Cruiser Squadron was strict.

At about this time the squadron legal officer appears to have prepared for Adm. Caperton the following untitled and undated memorandum:

"Our duty in protecting American interests is partly covered by [U.S. Navy] Regulations, Articles 1646 – 1647 – 1648 – 1649 – 1650.

"Under these articles we are duty bound to confer with the diplomatic

21 USS Washington and USS Eagle deck logs, July 1915.

Consular representatives in all cases where injury to the United States or to its citizens thereof is committed or threatened, in violation to the principles of international law or treaty rights.

"The only treaties now existing between the United States and Haiti are: (a) a treaty of citizenship; (b) a treaty of extradition and (c) a treaty of arbitration.

"We have a right to land a force only on our initiative when the rights of self-preservation are involved. In the case of the United States it includes the protection of the State, its honor, and its possessions; that is, for example, the Consulate and its records and the lives and property of United States citizens against the arbitrary violence, actual or impending, whereby the State or its citizens may suffer irreparable injury.

"If we do land a force for this purpose, it must only be used as a last resort and then only to the extent which is absolutely necessary to accomplish the end required.

"In all cases where forces are landed, especially where the local authorities are unable to give adequate protection to life and property, the assent of such authorities, or some one of them, shall first be obtained, if it can be done without prejudice to the interests involved."[22]

In view of the circumstances surrounding Adm. Caperton's later armed landing at Port-au-Prince, it is important to note that his sole authority for taking the action was drawn from the general instructions contained in Navy regulations. Years later, he affirmed this in testimony before a Senate select committee investigating the intervention. His own operational plan for the Caribbean deployment contains the statement that, "the Commander of the Cruiser Squadron had no orders except to proceed to San Domingo City, Port-au-Prince, Havana and the east coast of Mexico for the purpose of acquainting himself with the political situation in these countries."[23]

His frequent complaint in official communications to the Navy Department during operations in Haitian waters was the silence of the department concerning U.S. policy toward Haiti. The admiral's use of the forces at his

22 U.S. National Archives, Naval Records Collection, Subject File Haiti, 1911-1927.

23 Ibid.

disposal, though no doubt influenced by discussions with U.S. diplomatic and consular officers, accordingly was predicated largely on his professional judgment and interpretation of his duty to safeguard American lives and property. Such latitude, extending even to armed intervention in the affairs of a foreign state, is almost inconceivable in the context of modern-day communications, contingency planning and strict civilian control of the military.

Adm. Caperton's attempt to apply "Marquis of Queensberry" rules to Haitian revolutionary activity, relatively successful in the earlier Sam revolution and moderating Bobo's terrorism at Cap Haitian, failed to forestall the savage climax of Haiti's era of ephemeral governments. This was played out in Port-au-Prince July 27-28, 1915 at a rare moment when no foreign warships were present in the harbor.

The following narrative has been pieced together from several contemporary sources, but draws principally on an official report filed by U.S. Chargé d'Affaires R. B. Davis, who was present in Port-au-Prince when the Vilbrun Sam government fell. [24]

From what has been written thus far about Haitian revolutions, it may be seen that they ordinarily followed a well-defined pattern with certain variations brought about by the unexpected death in office of a president. An "army" of dissidents would gather in a place remote from the capital, usually in the north, and after an almost leisurely advance to Port-au-Prince, arrive at the seat of government to find the old administration embarked or in diplomatic asylum. The revolutionary army's general would then be proclaimed president of the republic.

This was the sequence Adm. Caperton expected the Bobo revolution to follow, but early in the morning of July 27 a group of 50 or more asylum seekers led by Charles de Delva left their havens in foreign legations and attacked the presidential palace, having first secured control of machine guns commanding the grounds. This they apparently did by enrolling, over a period of several months, sympathizers in the palace guard. Supporters of President Sam were routed, with approximately 50 killed and 200 wounded.

24 R.B. Davis Jr. to Secretary of State, Memorandum, January 12, 1916.

President Sam and several of his officers barricaded themselves in the palace, but it soon became evident the attackers were intent on burning them out. Several smaller buildings in the complex were put to the torch. Oil was sent for to use in setting fire to the palace itself.

Taking advantage of heavy smoke from the burning buildings and temporary confusion in the revolutionary ranks, President Sam made a sudden dash for an iron gate in the wall that separated the palace grounds from the French Legation.

With admirable presence of mind he brought with him a large key, to open the gate, but the lock had long since rusted shut. His would-be assassins hard on his heels, he scaled the 20-foot wall in a burst of superhuman energy. At the top of the wall a shot from his pursuers struck him in the leg and he fell into the French Legation grounds. Charles Zamor, who for months had taken refuge there from Sam's police, helped him into the legation.

Meanwhile at the National Prison a wholesale slaughter of 167 political prisoners had been carried out, apparently on President Sam's prior instructions to the prison commander, Gen. Oscar Etienne, should he (Sam) be driven from the palace. Among those hacked to death was ex-President Oreste Zamor, the brother of Charles Zamor. The prisoners, almost without exception, were members of elite Port-au-Prince families.

When news of what happened leaked out, the streets filled with enraged and grief-stricken people. Gen. Etienne took refuge in the Dominican Legation, and was later found there by the father of three sons slain at the prison. Etienne was dragged into the street, shot, mutilated and his left in a gutter outside the legation.

So intense was the feeling of the people against him that no one would carry his body away for burial. It lay where it fell, to be spat on by passers-by until the following day when it was soaked with fuel and burned on the spot.

An enormous crowd of hysterical people gathered in the vicinity of the National Prison. Everywhere one looked small processions were seen led by two men carrying on a plank a body of one of the victims, followed by friends and relatives whose curses against the president mingled with those of the people who stood in the streets and watched. It is not clear just how it became generally known that the prison massacre was the result of an order given by President Sam but there seems to have been conviction in everyone's mind that he was personally responsible for it.

French Minister Girard sent urgent requests to the American and British chargés d'affairs that they come to the French Legation and render such assistance as they could should an attempt be made to remove President Sam. Minister Girard was gravely concerned for the safety of his wife and two young daughters, who were with him inside the legation.

Both chargés sent dispatches requesting the presence in Port-au-Prince of warships. Then they went to the French Legation, where they found Sam, who was with his wife and several of their small children, the ministers of finance and foreign relations, a few army officers and numerous servants.

President Sam was in a frenzy, creeping about the house like a hunted animal, so terror stricken that when he passed an open window he crawled on all fours out of fear that someone on the outside would shoot him. During the afternoon disorganized mobs entered the legation grounds on several occasions. They were dispersed with great difficulty. Toward evening the street outside the legation was nearly deserted.

Just before nightfall a friend of the two Haitian ministers seeking refuge in the legation came, and after some conversation all three men departed. That the ministers chose the hazards of passing through city streets rather than the safety of the legation convinced U.S. Chargé d'Affairs R.B. Davis there was a plan afoot to remove the president. Davis and his British colleague agreed to stay in the legation throughout the night.

Despite their apprehension, the night passed uneventfully. Through midmorning of the next day, July 28, calm prevailed in the city. Things seemed so quiet that Davis decided to risk a quick visit to his legation to see if a reply to the dispatch of the day before had been received. The British chargé agreed to await his return before making a similar check. About an hour after Davis left, however, and convinced that the continued quiet in the city made his presence unnecessary, he too departed.

In the meantime a doctor had been engaged to treat the president's leg wound, which was not serious. The wound was dressed and treated with a liberal application of pungent iodoform. Sam then returned to his hiding place, the French minister's private bathroom, which led off the second floor master bedroom. The door to the bathroom was concealed at Madam Girard's suggestion by sliding in front of it the minister's huge, high-backed bed.

The Girards then retired to the main sitting room, which ran completely through, from front to back, the large house. Both doors to the sitting room were unbarred and open, so that the front and back gates to the grounds, at the end of long gravel driveways, could be watched.

Unbeknownst to those inside the house, a group of about 80 men had entered an adjoining property and scaled the wall. They were on the porch in an instant and despite frantic efforts by the Girards to bar the doors, the men forced their way in and demanded that President Sam be surrendered. Girard courageously refused. A quick search of the house failed to uncover Sam's hiding place. Again his surrender was demanded, and again it was refused.

A second and more thorough search began. It is possible this too might have failed were it not for the strong odor of iodoform in the minister's bedroom. The bed was shoved aside, the heretofore concealed door broken down, and Sam discovered. He was slashed several times in the face, then dragged by his heels down the stairs and out of the house before the horrified eyes of his wife and children.

Loudly screaming his innocence of any connection with the prison massacre, he was kicked and shoved the length of the front driveway. At one point he hooked an arm through the spokes of a parked carriage. A blow from a club broke his arm and he was forced nearer the gate, beyond which he saw a raging, chanting mob.

The gate, of iron and spiked at the top, was locked. Sam's abductors attempted to throw him over it. His body caught, partially impaled, on the spikes at the top. From there he was clawed to the ground outside, in a gathering crescendo of screams.

U.S. Chargé d'Affairs Davis, meanwhile, had reached the American Legation where he read a telegram telling him that the USS *Washington* would arrive in Port-au-Prince that morning. In fact he already could make out the lines of a warship on the horizon as it prepared to enter the harbor. As he retraced his path to the French Legation, he heard a terrific howl of fury from that direction. He quickened his pace fearing something terrible was about to happen. As he turned into the street fronting the French Legation, his way was blocked by a mass of frenzied people completely filling the street. He could see that something or somebody was on the ground in the center of the crowd in front of the gate. A moment later a man disentangled himself and ran by him, with a severed hand from which the blood was dripping, the thumb of which he had stuck in his mouth. Behind that man came other men, with the feet, the other hand, the head and other parts of the fallen president displayed on poles. Each part was followed by a mob of screaming men and women. The trunk of Sam's mutilated body came last, dragged through the street at the end of a rope.

This was the procession viewed in part by Adm. Caperton from the bridge of the *Washington* as it dropped anchor in the harbor of Port-au-Prince.

PART TWO:

INTERVENTION AND OCCUPATION

In searching history one finds stories of small countries that more powerful nations have overcome only after practically annihilating them. This task is long and difficult, and in the end results in a fatal tax on the conquerors. In view of this it would be prudent not to undertake such a task in the 20th century, and under the eyes of the civilized world. And not only because this task today as well as yesterday is long, difficult and in the end of fatal cost to the nation attempting it but also because now more than ever it would be ignoble, base and criminal.

— Rosalvo Bobo, graduate in law and medicine at Paris and Haiti's Minister of the Interior, Port-au-Prince, December 1914.

7

THE ESTABLISHMENT OF AMERICAN CONTROL

For foreigners the situation was extremely critical. The foreign legations, up to this time, had never been invaded, but the invasion of the Dominican Legation, where General Oscar had taken refuge, followed the next day by the invasion of the French, took away the last safeguard against violence from the natives. The ignorant classes, having seen these legations violated with apparently no evil results to the invaders, could not be expected to respect their inviolability as they had in the past; and as a result there remained no place of safety, nor was there any Haitian authority which could be looked to for protection.

Robert Beale Davis Jr., U.S. Chargé d' Affairs, Haiti, July 1915.

The *Washington* anchored shortly before noon on July 28, 1915, and as soon as a boat could be lowered an armed party went ashore to escort the French minister and the American and British chargés for a hurried meeting with Adm. Caperton on board his flagship. The diplomats urged an immediate landing of an American force to restore some degree of sanity in the chaotic Haitian capital and to safeguard foreign interests. The French minister was particularly worried, stating his fear for his wife and daughters, alone and defenseless in the French Legation.

At the conclusion of this meeting, Adm. Caperton was convinced of the necessity for landing forces without delay, though he still took time to formally notify Gen. Polynice and several other members of a hastily organized Committee of Public Safety. Polynice, it may be stated in passing, was the executioner of Gen. Oscar Etienne, commander of the National Prison.

The French minister and the two chargés left the ship at 4:10 p.m. Perhaps as much for the desired effect ashore as for protocol, a 17-gun salute was fired in honor of the senior diplomat.

At 4:50 p.m. *Washington's* landing party consisting of two companies of Marines and three companies of seamen, all under the command of Marine Capt. George Van Orden was landed at Bizoton, a defunct naval yard about a mile southeast of the center of Port-au-Prince. The total strength of the landing party was 340 officers and men. The Marines had been sent ashore on numerous earlier occasions, both in Haiti and in Mexico, but the blue-jacket battalion had never before been landed either for duty or drill.

Washington's guns were manned during the landing, but fire support was not called for by Capt. Van Orden. He split his landing force into two parallel columns and marched into the center of the city, disarming all Haitian soldiers encountered along the way. Scattered shots were exchanged and two Haitians were killed and 10 wounded. There were no American casualties. A considerable amount of arms and ammunition was seized. Guards were posted at the foreign legations. At nightfall, bivouacs were established. Sanitary conditions at the principal bivouac location, the central market, were grim, as noted in a report made to the Bureau of Medicine and Surgery by a medical officer who accompanied the landing party:

"... Against the protest of the Medical Officer, the Seamen's Battalion and the ship's company of Marines were made to bivouac in the filthiest and most unsanitary place in the city. The floor of the marketplace, a square building roofed-over but open at the sides except for a low brick wall, was covered with filth; trituarted dried human excrement; dust fouled with spit; fruit skins; banana stems; and in one place a shallow pool of foul-smelling water was present. The whole place was pervaded with a heavy odor, worse than a stable, and more like an open-air privy. The posts, inside, which supported the roof, smelt heavily of dried urine. (It is the habit of many Haitians to urinate and defecate wherever they find a convenient place.) It was in this place the men were obliged to stay overnight and, in fact, until the middle of the following afternoon. The sailors, of course, had no cots to lie on. Cots for the Marines arrived at 2 a.m. The majority of the men got no sleep that night, preferring to sit on the curb or against the wall rather than lie in the filth. In this place there were absolutely no facilities for the men to police themselves. There was no urinal or privy accessible in the vicinity, neither was there any water obtainable for any purpose.

"Water from the ship arrived at 5 a.m. The distribution of it was not properly supervised by a petty-officer. Breakers were rolled on the dirty floor and subsequently opened with only a mechanical cleansing around the bungholes. The hospital corps endeavored to supervise further handling of the breakers, propping them off the floor, and sterilizing around the bungholes with carbolic acid...."[1]

Sanitation, the landing party was quick to learn, was not a Haitian long suit. The city water system, of French colonial vintage, was contaminated at hundreds of places where water mains had been broken into to form crude public "wells," which then received the plentiful effluvia from the streets. Pigs, chickens and ducks roamed the city almost at will, and a foul-smelling gutter in front of the U.S. Legation was a favorite wallowing place for hogs nominally the property of the Haitian President. The water in the harbor istself was so bad that most ships' captains refused to use it even for swabbing decks.

Sanitary conditions ashore were not the only problems faced by the landing party. Their equipment was grossly inadequate. Packs (quickly dubbed "mankillers") cut into shoulders. Sailors shoes were not designed for long marches, and their uniforms (sailor whites dyed brown) did not hold up well under rough use. The landing force suffered from insects and crawling vermin. The most regrettable weakness was lack of training for the duties they were assigned ashore. This was underscored the night of Jan. 29 when two members of the bluejacket battalion were gunned down in the street outside their bivouac by their own shipmates firing wildly in the dark.

The arrival of additional Marines permitted the gradual withdrawal of the sailors and their return to more customary duties afloat. The USS *Jason* brought a company of Marines from Guantanamo on July 29, the USS *Connecticut* landed the Second Marine Regiment on Aug. 4, and USS *Tennessee* arrived with the First Marine Regiment on Aug. 15. This brought the total

1 P.A. Surgeon R. Cuthbertson to Bureau of Medicine and Surgery, undated report, August 1915 (?).

strength in Haiti to 88 officers and 1,191 enlisted. The last of *Washington's* bluejacket battalion returned to the ship on Aug. 18.

Throughout the initial period of the occupation there was little fighting. Adm. Caperton at first had the grudging support of the Haitian Committee of Public Safety. It was pledged to the candidacy of Dr. Bobo. who as leader of the recent revolution, was thought to be President Sam's logical successor. The admiral's chief of staff, Capt. Edward L. Beach Sr., who represented Caperton's political negotiations ashore, was not impressed with Bobo's qualifications for high office. He passed his observations to his admiral who in turn sent them on to Washington where they eventually reached President Wilson in virtually the same language couched by Capt. Beach.

Bobo was pictured as an inflammatory and perhaps deranged self-seeker, with decided anti-American sentiment. This assessment of Bobo was not kept a secret by the admiral or his chief of staff. It led to an open break with Bobo's supporters on the Committee of Public Safety.

The landing of American forces was decided upon by roughly 4 p.m. on July 28, 1915, and the landing itself occurred less than an hour later. Adm. Caperton informed the Navy Department of his action immediately, but orders from Washington to land were not received until almost six hours after the American intervention had begun. There is little doubt though that President Wilson had already decided on that course of action, and that U.S. policy was to intervene at the first favorable opportunity.

Special envoy Paul Fuller's report to President Wilson concerning his unsuccessful attempt to negotiate a treaty was specific in its recommendation to land Marines for the purpose of establishing an honest and efficient Haitian administration. He viewed that as a necessary first step in securing a treaty, and stated his belief that only an American intervention could bring about a permanent change in Haiti's intolerable political condition.[2] The State Department's Division of Latin American Affairs fully concurred with Fuller's report. President Wilson's comment on the report was that "... It gives me a

2 Paul Fuller, Jr. to President Woodrow Wilson, Report, June 14, 1915.

great deal of concern. Action is evidently necessary and no doubt it would be a mistake to postpone [it for] long."[3]

Meanwhile, France applied decided diplomatic pressure for U.S. intervention. The French Embassy in Washington informed the Wilson administration that it was urgently desired that the United States take energetic action in Haiti; that if so, France would look with approval on the action and be willing to support it in every way provided that a just recognition of French claims in Haiti be made; that should the action taken not be energetic, but composed of half-way measures, France would not look on it with such approval.[4]

Despite what has been recounted above, Adm. Caperton's landing of forces in Haiti caught Wilson unprepared. The belated order from Washington to land stated that foreign interests should be protected and that the representatives of England and France were to be informed of that fact and asked not to land forces of their own. On July 29, Secretary of State Robert Lansing wrote to Secretary of the Navy Josephus Daniels that, "Under no circumstances should he (Caperton) withdraw American Marines from Haitian territory until I have had an opportunity of conferring with you."[5] For more than a week Caperton's requests for policy instructions went unanswered while, with a great deal of soul-searching, the policy was hammered out in Washington.

Advice to Wilson and his Secretary of State poured in from predictable sources. Gov. Fort suggested that the Haitian *cacos* (guerrillas) be disarmed through the purchase of their rifles for a dollar or two each. So great was the cacos' need for money now that the Americans had ended the revolutionary profession, he asserted, it seemed quite probable that within 30 days the government would be the owner of all the arms in Haiti if his suggestion were followed. Occupation authorities did in fact later pursue such a policy with marked success.

From James P. McDonald of the Haitian National Railroad came a long letter. He recommended a provisional president for Haiti who would then be

3 President Woodrow Wilson to Secretary of State, Memorandum, July 2, 1915.

4 Internal Memorandum of Division of Latin American Affairs, Feb. 26, 1927.

5 Secretary of State to Secretary of the Navy, Memorandum, July 30, 1915.

Capt. Edward L. Beach, Sr.

charged with the negotiation of a treaty giving the United States control of Haitian finances and the right to employ its armed forces to still rebellion. After the signing of the treaty, a permanent president would be chosen to replace the provisional one who, after his role in the making of the treaty, would be so unpopular that he no longer could be effective.

R.L. Farnham of the National City Bank in New York hurried to Washington to confer with Secretary Lansing who, unlike his predecessor Williams Jennings Bryan, was not so enamored of Farnham's reasoned argument or persuaded of the cleanness of his hands when it came to Haiti.

Of all those who had President Wilson's ear, either directly or indirectly through Lansing in the days immediately following U.S. armed intervention, Paul Fuller Jr. appears to have been most influential. He argued that all governmental functions in Haiti should be taken over by the United States and overseen by the War Department, pending an honest election and the establishment of a responsible Haitian government.

The legality of the American intervention in Haiti, under the U.S. Constitution, worried President Wilson. The State Department's Division of Latin American Affairs, quoting liberally from a 1912 memorandum prepared by J. Reuben Clark, gave its view on the subject. Clark, incidentally, was destined to be the author of a more famous memorandum in 1928 expressing a philosophy divergent from that which follows:

"Inasmuch as the use in foreign countries of the military forces of the government for the purpose of protecting American life and property therein situated does not amount to an act of war or to a declaration of war, it is doubtful if Congress has authority directly to control such use....

"It appears unnecessary to elaborate any argument going to show that the President as Chief Executive is, under the Constitution, primarily charged with the conduct of our foreign relations, including the protection of the lives and property of our citizens abroad, save insofar as the Constitution expressly vests a part of these functions in some other branch of government (for example, the participation of the Senate in the making of treaties)....

"Under the general right to protect American citizens, their property etc., it is thought that practical intervention does not constitute an act of war, but purposes to render to Haiti friendly services, or acts not amounting to war. If this intervention can accomplish the re-establishment of a government which can maintain peace in Haiti and can leave in the hands of the United States Government or of American citizens the financial control, that is, in the form of collection of customs and the disbursement of public funds received from all sources of national income, it is thought that such a reorganization would bring to Haiti that stability which is indispensable to its proper future development."[6]

It stretches the imagination to conceive "practical intervention" carried to the extent of achieving complete financial control as not constituting an act of war. Further, justifying the intervention on the grounds that it was necessary for the protection of American lives and property is difficult because throughout Haiti's miseries American lives and property were remarkably well respected. But that respect almost always was due to the threat of heavy retaliation, a threat quietly imposed by the near-constant presence of U.S. warships in Haitian waters.

Lansing, in spite of the solicited and unsolicited advice showered upon him, found the situation in Haiti both distressing and perplexing. There seemed to be no excuse, no grounds, for reprisal as there had been at Vera Cruz for the continued presence of Americans forces ashore.

The justification eventually accepted and offered to President Wilson was that it was the humane duty of the occupying forces to furnish the means to

6 Latin American Affairs Division to Secretary of State, Memorandum, Aug. 5, 1915.

relieve famine conditions in Port-au-Prince and other Haitian cities. If naval authorities were directed to take over the collection of customs on imports and exports, funds realized might be expended for the relief of starving people.

Conditions were grim in Haiti as a result of years of revolutionary disruption. Adm. Caperton reported that women and children were actually dying from hunger in the streets of Port-au-Prince. He requested relief funds from the Navy Department and received from Secretary of the Navy Daniels the terse direction to "inform influential Haitians no funds are available for supplying food. Suggest they form organized relief society and if possible devise some scheme for work. If necessary you are authorized furnish what provisions you can spare at Navy rates. Great danger of encouraging idleness should be kept continually in mind."[7]

There were, of course, no influential Haitians with the slightest interest in forming societies for the relief of starving masses.

On Aug. 4, 1915, President Wilson, with misgivings, decided on what to do in Haiti. In a memorandum to the Secretary of State he wrote:

"My dear Mr. Secretary:

"These are serious matters and my own judgment is as much perplexed as yours. I fear we have not the legal authority to do what we apparently ought to do, and that if we did do what is necessary, it would constitute a case very like that of Mr. Roosevelt's action in Santo Domingo, and have very much the same issue.

"I suppose there is nothing for it but to take the bull by the horns and restore order. A long program ... involves legislation and the cooperation of the Senate in treaty making, and must therefore await the session of our Congress.

"In the meantime this is plain to me:

"1. We must send to Port-au-Prince a force sufficient to absolutely control the city not only but also the country immediately about it from which it draws its food. I would be obliged if you would ascertain from the Secretary of the Navy whether he has such a force available that can reach there soon. (This

7 Secretary of the Navy to Commander Cruiser Squadron, U.S. Atlantic Fleet. Despatch, August 3, 1915.

will probably involve making the city authorities virtually subordinate to our commander. They may hand the city government over to us voluntarily.)

"2. We must let the present Congress know that we will protect it but that we will not recognize any action on its part which does not put men in charge of affairs whom we can trust to handle and put an end to revolution.

"3. We must give all who now have authority there or who desire to have it, or who think they have it, or are about to have it to understand that we shall take steps to prevent the payment of debts contracted to finance revolutions; in other words, that we consider it our duty to insist upon constitutional government there and will, if necessary (that is, if they force us to it as the only way) take charge of elections and see that a real government is elected which we can support...."[8]

Upon receipt of the above, Secretary Lansing drafted detailed instructions to Adm. Caperton and Chargé d'Affairs Davis to implement President Wilson's instructions.

The Naval General Board reaffirmed its position of a year earlier concerning Mole St. Nicholas, and Lansing included in the draft of his instructions to Caperton a statement to the effect that the United States disavowed any intention to seek the cession of it for a naval base. But President Wilson deleted this statement. He substituted words indicating that the question concerning cession of Mole St. Nicholas would be taken up at a later date with a reorganized Haitian government.[9]

It's not likely Wilson entertained any desire to acquire Haitian territory, though this was the interpretation of his wording by later critics of the intervention. Almost certainly he was simply expressing Yankee shrewdness by not giving away a valuable card in advance of treaty negotiations.

Meanwhile, Adm. Caperton and his chief of staff, Capt. Beach, were caught up in the whirlpool of Haitian politics. There were two major factions hoping to name President Sam's successor. The Haitian Committee of Public Safety, representing the most recent revolutionists, backed Dr. Bobo. The Haitian

8 President Woodrow Wilson to Secretary of State, Aug. 9, 1915.

9 Woodrow Wilson to Secretary of State, Aug. 9, 1915.

Haitian President Dartiguenave and Adm. Caperton.

Congress, then in session, leaned toward Sudre Dartiguenave, president of the Senate. But for the presence of American forces, it is likely events would have taken their usual course in Haiti and Bobo would have been elected handily.

To be sure, the Haitian Congress elected the president, but it had never before stood in the way of a general whose cacos were in the streets of Port-au-Prince. Bobo's open anti-Americanism made him quite unacceptable to both Adm. Caperton and Chargé d'Affairs Davis. Both sent uncomplimentary reports concerning his character to their superiors in Washington.

Capt. Beach held frequent meetings ashore with representatives of Haitian factions. He gradually became convinced that Dartiguenave was a reasonably honest politician. It seems likely that ex-U.S. Minister Furniss helped persuade him of that. Several meetings with Dartiguenave took place in Furniss' home. The same dispatches that damned Bobo soon began to praise Dartiguenave, especially in view of the latter's willingness to grant the United States everything it conceivably could want in a future settlement, including customs control and Mole St. Nicholas.

On Aug. 6, 1915, Adm. Caperton ordered that all Haitian soldiers not living in Port-au-Prince leave the city. Within hours, U.S. Marines began arresting those who did not comply. There was some resistance and two Haitians were killed. This move, made possible by the reinforcement of the landing party by the Second Marine Regiment, broke the back of the Bobo faction.

The next day Adm. Caperton received from the Navy Department its first statement of policy. He was directed to "conciliate" the Haitian people and assure them that the United States had no intention to keep its forces in Haiti any longer than necessary to effect the re-establishment of a stable government. This was followed on Aug. 9 by an extraordinary dispatch that stated, among other things, "whenever the Haitians wish, you may permit the election of a president to take place. The election of Dartiguenave is preferred by the United States."[10]

More precise instructions were sent to Chargé d'Affairs Davis, who was directed to confer with Caperton and, in a manner as best determined by the two of them, ensure that the following was made clear to the leading Haitian politicians:

"First: Let Congress understand that the Government of the United States intends to uphold it, but it cannot recognize action which does not establish in charge of Haitian affairs those whose abilities and dispositions give assurances of putting an end to factional disorders.

"Second: In order that no misunderstanding can possibly occur after election, it should be made perfectly clear to candidates as soon as possible and in advance of their election, that the United States expects to be entrusted with the practical control of the customs and such financial control over the affairs of the Republic of Haiti as the United States may deem necessary for an efficient administration. The Government of the United States considers it its duty to support a constitutional Government. It means to assist in the establishing of such a Government, and to support it as long as necessity may

10 Acting Secretary of Navy to Commander Cruiser Squadron U.S. Atlantic Fleet. Despatch, Aug. 9, 1915.

require. It has no design upon the political or territorial integrity of Haiti; on the contrary, what has been done, as well as what will be done, is conceived in an effort to aid the people of Haiti in establishing a stable Government and in maintaining domestic peace throughout the Republic."[11]

The above was passed almost verbatim by Capt. Beach to all Haiti's senators and deputies at an informal meeting early on Aug.11. Davis was present at this meeting and further explained the demands that would be made of the reorganized Haitian government by the United States. Later that day, Haiti's Committee of Public Safety, in a last effort to prevent the election of Dartiguenave, ordered Congress dissolved and sent soldiers to seal the doors to the chambers. This was anticipated, however, and U.S. Marines shooed the soldiers away. Captain Beach then confronted the committee and ordered it dissolved, stating that its further continuance would cause its members to be considered public enemies and rebels against the United States.

On Aug. 12, with Capt. Beach on the floor of the Haitian chambers, Sudre Dartiguenave was elected President of Haiti with 94 of the 116 votes cast. Immediately after the votes were tallied, enthusiastic cheering broke out among the delegates. The president-elect made a brief speech in which he thanked the American forces for their presence. When he left the chambers for the ceremonial ride to the palace, Capt. Beach rode with him in the presidential carriage.

Of conditions in the Haitian capital that day, Beach wrote: "Everything quiet. Port-au-Prince had the appearance of being owned in fee simple by U.S. Marines."[12] But the weather observation for the day, as recorded in the deck log of the USS *Washington*, was "stormy."

11 Secretary of State to R. B. Davis. Despatch, Aug. 10, 1915.

12 Captain Edward L. Beach Sr. Undated and untitled manuscript.

8

THE TREATY

The strongest is never strong enough to be always the master, unless he transforms strength into right, and obedience into duty.

— Jean Jacques Rousseau, "The Social Contract."

For the first time in more years than people then living could remember, Haiti had a president who did not owe his election to a revolutionary faction. To be sure, the election of Dartiguenave could by no stretch of the imagination be called democratic. He was, in a manner of speaking, a creature of the U.S. military intervention. However, depending upon the wisdom and justice of the Americans, his election brought hope to a people who for many generations seemed to have lived without it.

On election day in August 1915, an American military government was set up at Cap Haitian, and in the next few days the towns of St. Marc, Pétionville and Léogane were occupied. Capt. Beach, Adm. Caperton's exulting chief of staff, wrote:

Hark, hark, the dogs do bark,
The Cacos are leaving the towns.
All in rags and some in tags,
But none in velvet gowns.[1]

On Aug. 17 Capt. Beach formally introduced the U.S. chargé d'affairs to President Dartiguenave, whose government had not yet been recognized by the United States. Chargé Davis delivered to the president a draft of the proposed treaty. He asked that the Haitian Congress pass a resolution without delay authorizing Dartiguenave to enter into negotiations.

The president reaffirmed his intentions to give the United States everything it de-

1 Capt. Edward L. Beach Sr. Undated and untitled manuscript.

sired saying it was his belief that the Congress would interpose no serious objections.

The very next day, however, the U.S. Secretary of the Navy directed Adm. Caperton to take charge of the Haitian customs service and to use the funds collected "for organizing and maintaining an efficient constabulary, for conducting such temporary public works as will afford immediate relief to the discharged soldiers and starving populace by giving them employment, and for the support of the Dartiguenave Government."[2] This action struck Haitians at a most sensitive point, and boded ill for a successful conclusion of treaty negotiations.

On Aug. 21 the customs house at Cap Haitian was taken over and by Sept. 2, with the seizure of customs collection at Port-au-Prince, the operation was complete. Navy paymasters, at first from ships of the cruiser squadron, and later by officers ordered from the United States, were placed in charge of every Haitian port as collectors of customs and captains of the port. Untrained and almost entirely unprepared for such duty, they brought to this unusual assignment a spirit of innovation, and an honest administration quickly transformed the Haitian customs service.

An "Admiral Caperton Account" was opened at the National Bank of Haiti. All funds collected were paid into it. Military commanders were then authorized to draw on it under two appropriations: "Public Works" and "Military and Civil Government of United States Forces." Great leeway was granted in administering the accounts. Relatively large sums were expended to purchase arms from the cacos (as Gov. Fort earlier had recommended to President Wilson).

One problem of long standing in the Haitian economy prior to the intervention was a wildly fluctuating foreign exchange rate. During the first months of the U.S. Navy's administration of the customs service, a fixed rate of five gourdes to the dollar was established. An officer who participated in the stabilization wrote that Haitians vigorously opposed this, according to French economists who warned that Navy paymasters were ignorant of economics and were trying to accomplish the impossible. The exchange rate was stabilized,

2 Secretary of the Navy to Commander Cruiser Squadron, U.S. Atlantic Fleet. Despatch, Aug. 18, 1915.

Franklin D. Roosevelt (left), Secretary of the Navy Josephus Daniels, (center).

however, and remained so. Perhaps the most touching argument against this U.S. action was that put forth by the Haitain minister of finance, who said his friends were accustomed to fluctuating rates, counted on them, and that it was a crime to ruin their businesses.[3]

When the United States took control of the customs service, Haiti's public debt was estimated as follows:

External debt, $20 million (market value, $16 million; internal debt, $2.25 million (market value, $1.625 million); floating debt (interest guarantees, etc.) $2.25 million. All of it was secured in one way or another by claims on customs revenue. It was established that the priority was a rapid retirement of public debt, a policy that likely prolonged slow or no growth of the Haitian economy.

3 Capt. Charles Conard (SC) USN, "A Year in Haiti's Customs and Fiscal Service." U.S. Naval Institute Proceedings. April 1923.

While U.S. paymasters were at work reorganizing the customs service, Adm. Caperton sought to wrap up the treaty. President Dartiguenave seemed agreeable from the start, but American takeover of customs in advance of negotiations led to sharp resistance to the treaty in Haiti's Congress. Under Haiti's Constitution, treaty ratification required approval by both the Haitian Chamber of Deputies and the Senate. Failure of Haiti to accept immediately U.S. proposals (which went far beyond those suggested by Paul Fuller prior to the intervention) was viewed by some American officials as rank ingratitude. Secretary of State Lansing expressed his misgivings in a note to President Wilson:

"I confess that this method of negotiation, with our Marines policing the Haitian capital, is high handed. It does not meet my sense of a nation's sovereign rights and is more or less an exercise of force and an invasion of Haitian independence. From a practical standpoint, however, I cannot but feel that it is the only thing to do if we intend to cure the anarchy and disorder which prevails in that Republic....

"I have not been unmindful of the possible criticism which may be aroused in the (U.S.) Senate in case this treaty should be signed and submitted to them for action. As I said, it seems a high-handed procedure, but I do not see how else we can obtain the desirable end of establishing a stable government in Haiti and maintaining domestic peace there."[4]

The U.S. President replied:

"This is, I think, necessary and has my approval. Do you think it will affect Latin American opinion unfavorably?"[5]

Measures taken to secure the treaty on American terms were extraordinary. On Aug. 25, Acting Secretary of the Navy Franklin D. Roosevelt, Secretary of State Lansing and Capt. Beach (temporarily recalled to the United States due to his wife's serious illness) met in Washington. They drafted a message signed by Beach to President Dartiguenave assuring him of continued support by the United States should he dissolve the Haitian Congress, dismiss his

4 Secretary of State to President Woodrow Wilson. Memorandum, Aug. 13, 1915.

5 Woodrow Wilson to Secretary of State. Memorandum, Aug. 13, 1915.

cabinet and form a new government of "patriots" willing to support the treaty.[6]

Meanwhile, the U.S. chargé d'affairs in Haiti was instructed to orally impress de facto Haitian authorities of the serious consequences that would attend rejection of U.S. demands. Should that happen he was to say that two courses of action would be considered. The United States would either establish a military government pending honest elections or permit control of the Haitian government to pass to those willing to cooperate with American authorities.[7]

But on Aug. 27 the Haitian government issued a written reply in which practically every stipulation in the proposed treaty was either rejected or so reworded as to defeat its purpose. The American Legation was then instructed to conclude the treaty as originally drafted without modification and to inform the Haitian government that it was expected to proceed without undue delay.[8]

Adm. Caperton applied more direct pressure on Sept. 3 by proclaiming martial law. The order applied to Port-au-Prince and all other parts of Haiti under his military control, which by then included virtually all important cities and towns. The proclamation was accompanied by a decree aimed at the Haitian press in which trial before military courts was established for "the publishing of false or incendiary propaganda against the Government of the United States, or the Government of Haiti, or the publishing of any false, indecent or obscene propaganda, letters signed or unsigned, or other matter which tends to disturb the public peace."[9]

President Dartiguenave reported that these measures had strengthened his hand, but that certain influential members of his Cabinet remained adamant against those portions of the proposal having to do with financial matters. Reassured of Caperton's continued support, he then demanded and received the resignations of his minister of public works and minister of foreign affairs.

American offensive operations against the cacos who, particularly in the north, were becoming troublesome, were held in abeyance while Dartiguenave

6 Acting Secretary of the Navy to USS Washington, Despatch, Aug. 24, 1915.

7 Secretary of State to American Lega*tion, Port-au*-Prince, Despatch, Aug. 24, 1915.

8 Secretary of State to American Legation, Port-au-Prince, Proclamation, Sept. 3, 1915.

9 Office of the Provost Marshal, Port-au-Prince, Proclamation, Sept. 3, 1915.

Haitian President Dartiguenave.

pressured his reorganized Cabinet to accept the treaty. Finally, on Sept. 16 the "Treaty Between the United States and Haiti — Finances, Economic Development and Tranquility of Haiti" was signed by new Minister of Foreign Affairs Louis Borno and Chargé d'Affairs Davis. Of its 16 articles the following are particularly noteworthy as they form the legal basis for much of what subsequently was attempted by the Occupation:

"Article I: The Government of the United States will, by its good offices, aid the Haitian Government in the proper and efficient development of its agricultural, mineral and commercial resources, and in the establishment of the finances of Haiti on a firm and solid basis.

"Article II: The President of Haiti shall appoint, upon nomination by the President of the United States, a general receiver and such aides and employees as may be necessary who shall collect, receive and apply all customs duties in imports and exports accruing at the several custom houses and ports of entry of the Republic of Haiti.

"Article IV: The President of Haiti shall appoint, upon nomination by the President of the United States, a financial adviser who shall be an officer attached to the ministry of finance, to give effect to whose proposals and labors the minister will lend efficient aid. The financial adviser shall devise an adequate system of public accounting, aid in increasing revenues and adjusting them to the expenses, inquire into the validity of the debts of the Republic, enlighten both Governments with reference to all eventual debts, recommend methods of collecting and applying the revenues, and make such other recommendations to the minister of finance as may be deemed necessary for the welfare and prosperity of Haiti.

"Article V: All sums collected and received by the general receiver shall be applied, first, to the payment of the salaries and allowances of the general receiver, his assistants and employees, and expenses of the receivership, including the salary and expenses of the financial adviser, which salaries will be determined by previous agreement; second, to the interest and sinking fund of the public debt of the Republic of Haiti; and, third, to the maintenance of the constabulary referred to in Article X, and then the remainder to the Haitian Government for purposes of current expenses....

"Article VIII: The Republic of Haiti shall not increase its public debt except by previous agreement with the President of the United States, and shall not contract any debt or assume any financial obligation unless the ordinary revenues of the republic available for that purpose, after defraying the expenses of the Government, shall be adequate to pay the interest and to provide a sinking fund for the final discharge of such debt.

"Article IX: The Republic of Haiti will not without a previous agreement with the President of the United States modify the customs duties in a manner to reduce revenue there-from; and in order that the revenues of the Republic may be adequate to meet the public debt and the expenses of the Government, to preserve tranquility and promote material prosperity, the Republic of Haiti will cooperate with the financial adviser in his recommendations for the improvement in the methods of collecting and disbursing the revenues and for new sources of needed income.

"Article X: The Haitian Government obligates itself, for the preservation of domestic peace, the security of individual rights, and full observance of the provisions of this treaty, to create without delay an efficient constabulary, urban and rural, composed of native Haitians. This constabulary shall be organized and officered by Americans appointed by the President of Haiti, upon nomination by the President of the United States. The Haitian government shall clothe these officers with the proper and necessary authority and uphold them in the performance of their functions. These officers will be replaced by Haitians as they, by examination, conducted under direction of a board to be selected by the senior American officer of this constabulary and in the presence of a representative of the Haitian Government, are found to be qualified to assume such duties. The constabulary herein provided for, shall, under the direction of the Haitian Government, have supervision and control of the arms and ammunition, military supplies and traffic therein, throughout the country. The high contracting parties agree that the stipulations in this article are necessary to prevent factional strife and disturbances.

"Article XI: The Government of Haiti agrees not to surrender any of the territory of the Republic of Haiti by sale, lease or otherwise, or jurisdiction over such territory, to any foreign government or power, nor to enter into any agreement or contract with any foreign power or powers that will impair or tend to impair the independence of Haiti.

"Article XII: The Haitian Government agrees to execute with the United States a protocol for the settlement, by arbitration or otherwise, of all pending pecuniary claims of foreign corporations, companies, citizens or subjects against Haiti.

"Article XIII: The Republic of Haiti, being desirous to further the development of its natural resources, agrees to undertake and execute such measures as in the opinion of the high contracting parties may be necessary for the sanitation and public improvement of the Republic, under the supervision and direction of an engineer or engineers, to be appointed by the President of Haiti upon nomination of the President of the United States and authorized for that purpose by the Government of Haiti.

"Article XIV: The high contracting parties shall have authority to take such steps as may be necessary to insure the complete attainment of any of the objects comprehended in this treaty and, should the necessity occur, the United States will lend an efficient aid for the preservation of Haitian independence and the maintenance of a government adequate for the protection of life, property and individual liberty.

"Article XVI: The present treaty shall remain in full force and virtue for the term of 10 years, to be counted from the day of exchange of ratifications, and further for another term of 10 years if, for specific reasons presented by either of the high contracting parties, the purpose of this treaty has not been fully accomplished....[10]

The treaty said nothing about the presence of American forces in Haiti, nor was anything said about American control or influence over the Haitian judiciary or public education. These apparent oversights provoked heated controversies during the Occupation. The signing of the treaty was quickly followed by formal diplomatic recognition by the U.S. government of the Dartiguenave government. At 8:55 a.m. on Sept. 17, 1915, Adm. Caperton's flagship fired a national salute of 21 guns with the Haitian ensign at the fore.

The prospective treaty signed was not the treaty ratified, however, and further American pressure was applied to bring the recalcitrant Haitian Congress into line. Funds for the ordinary expenses of the Haitian government, including salaries for senators and deputies, were frozen. On Oct. 3 Dartiguenave threatened to resign if the overdue salaries were not paid. He was told they would be — as soon as the treaty was ratified. The Haitian Chamber of Deputies gave its consent on Oct. 6 by a vote of 75 to 6. Adm. Caperton then authorized an allowance to be paid to the Haitian government to meet current expenses, but at the direction of the Navy Department, left open the question of paying salaries already owed to government employees, pending final ratification.

Opposition continued in the Haitian Senate, and the U.S. State Department became more and more apprehensive that the treaty would fail. This seemed a real possibility when, on Nov. 8, the Haitian Senate issued a lengthy report declaring certain articles of the treaty unconstitutional. It urged a renewal of negotiations with a view to obtaining terms that would virtually nullify U.S. control.

Further and more drastic measures evidently were needed to secure ratification. Secretary of the Navy Daniels, after conferring with Secretary of State Lansing, directed Adm. Caperton to make the following statement to President Dartiguenave and his Cabinet:

"I have the honor to inform the President of Haiti and the members of his Cabinet that I am personally gratified that public sentiment continues favor-

10 Treaty With Respect to the Finances, Economic Development and Tranquility of Haiti, U.S. Treaty Series, Number 623.

able to the treaty; that there is a strong demand from all classes for immediate ratification, and that the treaty will be ratified Thursday. I am sure that you gentlemen will understand my sentiment in this matter, and I am confident if the treaty fails on ratification that my Government has the intention to retain control in Haiti until the desired end is accomplished, and that it will forthwith proceed to the complete pacification of Haiti so as to insure internal tranquility necessary to such development of the country and its industry as will afford relief to the starving populace now unemployed.

"Meanwhile the present Government will be supported in the effort to secure stable conditions and lasting peace in Haiti, whereas those offering opposition can only expect such treatment as their conduct merits. The United States Government is particularly anxious for immediate ratification by the present Senate of this treaty, which was drawn up with the full intention of employing as many Haitians as possible to aid in giving effect to its provisions, so that suffering may be relieved at the earliest possible date. Rumors of bribery to defeat the treaty are rife, but are not believed. However, should they prove true, those who accept or give bribery will be vigorously prosecuted."[11]

In private instructions to the admiral, appended to the above, Secretary Daniels stated that it was expected that the American position would be made sufficiently clear to remove all opposition and to secure immediate ratification. That expectation was fulfilled, and on Nov. 12 the Haitian Senate approved ratification by a vote of 25 to 9. Pending approval by the U.S. Senate (which did so unanimously on Feb. 28, 1916), a modus vivendi was entered into by both governments that permitted the treaty terms to be effective immediately.

Thus did the United States secure a treaty and assume the "obligations" which would be cited in the years that followed as both justification and necessity for a prolonged and increasingly unpopular military occupation of the Black Republic.

11 Secretary of the Navy to Commander Cruiser Squadron, U.S. Atlantic Fleet. Despatch, Nov. 10, 1915.

9

PACIFICATION

We were all imbued with the fact that we were trustees of a huge estate that belonged to minors.... The Haitian people are divided into two classes, one class wears shoes and the other does not.... Ninety-nine percent of the people of Haiti are the most kindly, generous, hospitable, pleasure loving people I have ever known. They would not hurt anybody. They are most gentle when they are in their natural state. When the other one percent that wears vici kid shoes with long, pointed toes and celluloid collars, stirs them up and incites them with liquor and voodoo stuff, they are capable of the most horrible atrocities; they are cannibals. They ate the liver of one Marine....

Those that wear shoes I took as a joke. Without a sense of humor you could not live in Haiti, among the shoe class.... But the people who were barefooted, the women wearing Mother Hubbards and the men dungarees halfway up to their knees, with scarred feet, indicating the hardest kind of toil, and with great blisters on their hands, and the palms of their hands as hard as a piece of sole leather — those people you could absolutely trust. I went all over Haiti living with them in their shacks, and they always gave you the best they had — food and anything they had in the world. They did not know the value of anything. They did not know anything about time, distance or value.... What we wanted was clean little towns, with tidy thatch-roofed dwellings. That is what the country can afford, and what it ought to have....

— Brig. Gen. Smedley D. Butler, USMC, Oct. 27, 1921.

With the treaty ratified, Adm. Caperton turned his attention to settling affairs with the cacos of the north who, with increasing boldness, were interrupting the flow of food, water and goods for export to the principal towns and ports of a large part of the country. The cacos were not a numerous force, and earlier efforts to "buy" them and their weapons met with some

Brig. Gen. Smedley D. Butler, USMC.

success. As much as 50,000 gourdes were offered in September for the surrender of each 1,000 cacos and their weapons. An American guarantee that the Haitian government's proclamation of general amnesty would be respected was widely published throughout the country. Despite this, it soon became evident that a military campaign would have to be launched against diehard remnants of the caco armies.

In the period from Nov. 1-17, 1915, their strongholds in the mountains of the north were systematically reduced. Personal heroics on the part of Marines in this endeavor won them lasting fame in the Corps. No Marines were killed and only two were wounded in the entire campaign, the climax of which was the capture of Fort Riviere. Perhaps 100 cacos lost their lives. The news of what he considered an excessive shedding of Haitian blood caused Secretary of the Navy Josephus Daniels to order suspension of the American offensive.

The success of the U.S. military action in Haiti, where a relative handful of Marines disarmed and pacified a population of 2.5 million people dispersed over rugged and difficult terrain, seems incredible in view of the American failure in Vietnam. The explanation lies not only in the undoubted courage and professionalism of the Marines and their great superiority in weaponry but also in the nature of the cacos soldier who was typically half-starved and recruited to fight for a cause of which he knew little and cared less. It was not, as some suggested at the time, an inherent inferiority of the Haitian fighting spirit (the French experience during the Haitian Revolution disproved that), but almost entirely the failure of elite Haitian leadership to persuade the cacos soldier that he was fighting for anything other than his general's personal gain. Except perhaps for a very brief period in 1919, U.S. Marines, in the eyes of a large majority of Haitian peasants, were more welcomed than the cacos.

One of the first acts of the U.S. Occupation authorities was to disband the regular Haitian Army. Prior to the intervention, it consisted of 38 line and four artillery regiments, four regiments of the Presidential Guard, and a Gendarmerie of some 1,800 men. The paper strength of the Haitian Army and Gendarmerie was 11,000, officered by 308 generals and 50 colonels, plus additional honorary high-ranking officers appointed by the president from among friends and supporters. The pay of a private was one gourde a month, plus four gourdes for subsistence, little of which was ever received. This money usually wound up in the generals' pockets.

Under Article X of the treaty, an "efficient constabulary" was established to replace the armed forces. Even before the treaty was ratified, steps were taken in that direction. In September 1915 the first units of the Haitian Gendarmerie were formed with an initial strength of 336 men, most of them privates. By Feb. 1, 1916 the Gendarmerie was responsible for policing the entire country. Its strength had increased to 2,533 Haitians and 115 American commissioned and non-commissioned officers detailed to the Gendarmerie for service.

In June 1916 the U.S. Congress passed legislation authorizing Americans to serve in Haitian forces. U.S. Marines and naval personnel were commissioned in the Gendarmerie while still holding rank and entitled to pay from their parent services. Gen. Smedley D. Butler, who had received the Medal of Honor for his service at Fort Riviere, was commissioned a major general in the Gendarmerie and became its first commander. The strength and efficiency of the Gendarmerie d'Haiti progressed so rapidly that by the end of 1916 the Marine Brigade was withdrawn from the interior of the country to coastal cities and towns, principally Port-au-Prince and Cap Haitian.

Haitians joined voluntarily on enlistment papers similar to those used by U.S. servicemen. They wore Marine uniforms, with the exception of the Marine Corps emblem. They were issued Marine small arms and rifles, and received

Marine training. There was no difficulty in getting volunteers, although the physical condition of many recruits was appalling.

"We made every effort to cure those that we had, so that they would not go to sleep standing up in the daytime. That was one test. I have frequently found a sentry on a post in front of an important building sound asleep, standing up with the sun shining in his face. That is not his fault. He was diseased. An examination showed that 90 percent of them had blood diseases and that 85 percent had intestinal worms, and we took immediate measures to cure it, and before I left Haiti the gendarmes could keep awake for two or three hours.... Never during the time I was in Haiti, nor from any reports I have received since, has any disloyalty on the part of a gendarme occurred.... My orderly in Port-au-Prince was the worst caco in Haiti, and I picked him out because he was the ugliest brute I ever saw, and I trusted him with my children, my wife and everything. He was the most faithful man I have ever known.... Whenever I had an inspection to make in the woods, I left my family with this ugly caco sleeping on a couch on the front porch of my house, and he never would move for two weeks. They would take his food to him. And nobody would come into the yard either. I trusted him absolutely. It is a great army, that Gendarmerie...."[1]

Within the Gendarmerie, Haiti was divided for purposes of command into departments headed by colonels of the force, and further divided into districts. The Department of Port-au-Prince had seven districts, the Department of the Cap eight, and the Department of Cayes three. A Gendarmerie captain typically commanded districts. Districts were divided into three or four sub-districts commanded by lieutenants of the Gendarmerie. The police outposts were permanently established at strategic points and were commanded by non-commissioned Haitian officers. Two squadrons of cavalry were held in reserve at Port-au-Prince and at Cap Haitian. The Gendarmerie service is one of the bright spots of the Occupation and extended far beyond normal military and police duties.

1 Statement of Brig. Gen. Smedley D. Butler to Select Committee on Haiti and Santo Domingo, Oct. 27, 1921.

Commencing Jan. 1, 1916, all salaries and expenses of every nature pertaining to the Haitian government were paid directly to claimants by agents of the American Occupation. This raised a storm of protests from Haitian officials who saw an important source of their income — the discounting of *feulilles* (promissary notes), legitimate claims against the government — taken away. The system adopted was to effect payment by non-negotiable checks, thus assuring that funds actually reached the hands of those for whom they were intended. This "implied insult" to government officials was heightened by American action taken to verify the correctness of pay lists provided by the Haitian government.

Municipal elections in January 1916 drew the attention and concern of Occupation authorities. When the election lists at the city of Petitgoave mysteriously disappeared on the eve of elections, the mayor was arrested by the Gendarmerie and American officers moved in to supervise the balloting. Haitian government officials viewed this and other Occupation actions as exceeding that permitted by the treaty. Working relationships with Haitian officials steadily worsened, particularly when early promises of American aid and capital inflow failed to materialize.

An attempt to negotiate a $2 million loan from a New York bank failed when the bank insisted on a "moral guarantee by the United States Government," and such was not given. Thus Haiti's financial plight worsened in the early months of the intervention, despite American control of the customs. Interest as well as principal payments on the public debt went into default.

In fairness to U.S. authorities, however, this was the almost inevitable result of conditions brought about by the war in Europe, particularly the collapse of the French coffee market and a drastic decline in Haitian exports. The general economic condition and specific American actions to make the Haitian government more efficient led to widespread unemployment that fell hardest on those who formerly made their living on the labor of the peasant class or on non-productive government work.

Conditions in the cities, particularly Port-au-Prince, reached famine proportions in the first winter of the intervention, and U.S. efforts to provide relief were not overly generous.

It is interesting, in the light of the welfare ethic that would flower half a century later in America, to examine the modest program placed in effect by U.S. officials in Port-au-Prince. The following is taken from a report made to the Navy Department by P.A. Surgeon P.E. Garrison titled "Account of the Red Cross relief work in Port-au-Prince, Haiti, January to April 1916."

"For several months waste food collected from American naval vessels in the harbor of Port-au-Prince, and from Marine forces ashore, had been distributed at the Wesleyan Mission under the direction of its pastor, the Rev. Mr. Turnbull. This work has continued, and the cost of collection and distribution paid out of Red Cross funds (some $1,500 was donated by the Red Cross during this period). From 10 to 25 barrels of food have been collected and distributed daily.... Our records show how that from 250 to 400 persons were fed once or twice daily with this food, the average number in the 'bread line' during 13 weeks being 320....

"In as much as the daily amount of food collected from the ships varied greatly, according to the number and sizes of ships in the harbor, a certain supply of dry provisions was kept on hand to supplement the distribution on poor days....

"These dry groceries were used also to relieve a form of poverty which the 'bread line' could not reach. Our attention was called repeatedly to cases of extreme need in families, of educated and self-respecting people who, from one cause or another, had been reduced to a condition of actual want. The chief cause active in producing such conditions were loss of employment occasioned by recent political changes, the failure of the Haitian government to pay salaries for several months, and the depressed state of business and industry.... Such families would not stand in line with a ragged and dirty crowd of humanity clamoring for a portion of the food collected from the ships, but they would come quietly to the depot with their identification cards and receive a small package of rice or beans to be cooked or eaten at home....

"We have already drawn attention to the danger of encouraging and prolonging a state of indigence, especially among an ignorant population whose sense of self-respect is not sufficiently developed to make dependence obnoxious. The number of persons applying to the Red Cross for food constantly tends to increase, and undoubtedly would increase indefinitely if not controlled. The bread line becomes a regular point of call for the professional beggars of the city. Many not really in need come with false stories of distress. Two or more members of the same family have been discovered, each separately drawing food for the entire family. A number have attempted to draw a double portion of food by applying at the two depots under different names. A few instances have been reported to us of families leaving their homes and coming long distances to Port-au-Prince because they heard that food was being given away here. Two extreme instances have come to our attention of men declining opportunities to work for wages because the work would interfere with their daily calls at the food depot...."

On Jan. 1, 1916, the USS *Washington*, which had lain at anchor in the Port-au-Prince harbor for six months, her crew for the greater part of this time confined to the ship except for "recreational parties" permitted in late-October and thereafter, got under way for Guantanamo and eventually the United States. A thousand little dramas are recorded in the ship's log from the hot, sultry days and nights spent in Haiti. The ship's bugler lost three front teeth in a fight on board and his injury was logged as "not in the line of duty." A wardroom cook named Ah Hee was taken to the Captain's Mast for "gross negligence, allowing soap to get into the wardroom lunch." Two firemen were warned for "cheering the French cruiser," and an ordinary seaman returning from one of the recreation parties ashore was sentenced to loss of pay in the amount of $3.48 for "failing to report to the venereal head for prophylactic treatment after having been lawfully directed to dose."

The flag of the cruiser squadron commander was shifted to USS *Tennessee* upon *Washington's* departure. The commanding officers of these two ships relieved each other so that Capt. Beach of the *Washington* could continue to act as Adm. Caperton's chief of staff.

An increasingly bitter relationship between the Dartiguenave administration and the Haitian Congress climaxed on April 5, 1916, by the dissolution of the Senate, which had threatened to impeach the president. Dartiguenave's action had the reluctant support of Adm. Caperton. The president tried to get the Haitian Chamber of Deputies to function as a constituent assembly, a project of doubtful legality, but the chamber refused the constitutional changes.

It appeared that such would have to await the election of a new legislature, scheduled for the following January.

On May 10, 1916, disturbances in the Dominican Republic called Adm. Caperton to Santo Domingo, where he was to remain until his relief by Rear Adm. Charles F. Pond on July 18, 1916. Military command in Haiti was exercised by the Marine Brigade commander, who reported to the Navy Department through Adm. Caperton in Santo Domingo.

By this time Navy and Marine officers had assumed administrative control not only of the customs service but also the departments of public safety, public works and public health. In June 1916 President Wilson nominated and President Dartinguenave appointed, under the terms of the treaty, the first general receiver. Navy paymasters were relieved of their duties in the Haitian customs service on Aug. 29. Thereafter the character of the Occupation regime gradually became more civilian than military. This did not, however, improve the Occupation's working relationship with the Haitian government. From the very start, strong disagreement arose between the minister of finance and his financial adviser. Despite their respective titles, there was little question where the decision-making power rested.

A year after the initial landing of U.S. forces in Haiti, Americans there looked with increasing disfavor on the Haitian government they had been instrumental in creating. They considered it not only hopelessly inefficient and insincere in carrying out the Occupation agreements but thoroughly detested by the Haitian people. The government remained in office only by the presence of armed American force.

The solution to Haiti's troubles seemed as far away as ever.

10

A NEW CONSTITUTION

I wrote the Haitian Constitution myself, and if I do say it, I think it's a pretty good constitution.

— Assistant Secretary of the Navy Franklin D. Roosevelt, Aug.18, 1920.

If I should be, as I fully expect to be, elected as President of this just and honorable Republic, I will not empower an Assistant Secretary of the Navy to draft a Constitution for helpless neighbors in the West Indies and jam it down their throats at the points of bayonets borne by United States Marines.

— Sen. Warren G. Harding, Aug. 28, 1920 .

As the year 1916 drew to a close, something in the nature of organized opposition to the Dartiguenave government arose in Port-au-Prince. The opposition was fairly divided into two groups calling themselves "The Progressives" and "The Nationalists."

The former claimed to be in favor of doing everything necessary for the development of the country, while attacking the president for being too easily controlled by the Americans. The latter agitated openly for black rule and charged that Dartiguenave was preparing to amend the Constitution in such a way as to grant white ownership of land, the implication being that such would be but the first step toward a return of slavery. Both groups were violently anti-Occupation and pledged to impeach the president.

Dartiguenave commanded little respect from American authorities in Haiti. Questions were raised with Washington concerning what should U.S. policy be if he faced ouster by the new legislature to be elected in January 1917. A Navy Department memorandum dated Jan. 6, 1917 stated that the United States government "will not countenance any impeachment proceedings which may be entered into by the legislative body of Haiti, but, on the contrary, will

Secretary of the Navy Josephus Daniels, left, Franklin D. Roosevelt right.

consider such action as a move to begin revolutionary and disorderly political activity in the Republic."

This policy was clearly explained to the Haitian government and to the political opposition by U.S. diplomatic and military representatives in Port-au-Prince. The reason for continuing U.S. support of Dartiguenave was in part his suddenly more cooperative attitude toward the Occupation, and more particularly the need for certain changes in the Haitian Constitution that it was thought could only be secured with the wholehearted backing of the president.

In November 1916 a draft of proposed constitutional changes was presented to Occupation authorities and to the U.S. Legation. It was forwarded to the State Department in Washington, and after a careful review certain alterations were "suggested." These included the elimination of a five-year residency requirement for foreigners to acquire real estate in Haiti (the complete text of this article in the new constitution eliminated a bar to white ownership that had existed for 113 years) and ratification of all acts by U.S. military authorities in the republic. These changes were considered most necessary.

Numerous others were given as being of lesser importance to the United States. On this list were curtailment of exorbitant expense accounts for secretaries of state; disestablishment of the Haitian Council of State, which was considered unnecessary in view of the existence of a legislative branch; impeachment of justices of the Haitian Supreme Court and the Court of Appeals in the same manner as impeachment of cabinet officers; nondiscrimination of foreigners in matters of personal protection; elimination of the Haitian Chamber of Accounts, which was considered redundant in view of the offices of collector of customs and the financial adviser.

With considerable difficulty, the Dartiguenave government was persuaded to accept all of the proposed changes in its draft constitution. Today, none of the changes appear to be unreasonable, and the only controversial point concerned foreign ownership of property, considered essential if foreign capital investments in Haiti were to occur, though for years the constitutional bar to ownership had been sidestepped by foreign investors, especially Germans, by the simple tactic of creating dummy Haitian partnerships.

In January 1917 the election of a new national assembly was held without incident. By all accounts it was as fair and honest as any election Haiti had ever known. Gendarmes supervised polling places, with their American officers kept discretely in the background. A date in April was set for the assembly to consider adoption of the new constitution.

In the latter part of January and the early weeks of February, Assistant Secretary of the Navy Franklin D. Roosevelt, accompanied by the Commandant of the Marine Corps, Major Gen. George Barnett, made an inspection trip to Haiti. Arriving in Port-au-Prince, they crossed the country to Cap Haitian on horseback. Along the way, Roosevelt talked freely with Haitian and Occupation officials. He apparently was much impressed by the progress being made in road building and sanitation work. He reported having heard almost no complaints directed at the Occupation.

Impending war with Germany cut short a planned excursion to the Dominican Republic, and Roosevelt's party hurried back to Washington from

Cap Haitian.

In March, the Haitian government formally requested the good offices of the United States in the negotiation of a $30 million loan to develop the country's resources and to take advantage of an opportunity to retire some of its foreign debt, an opportunity brought about by the war in Europe. In order that the proposed loan might be placed on advantageous terms, the U.S. government thought it desirable to extend the life of the treaty for 10 years, as permitted by Article XVI, thus assuring those who invested in Haitian bonds that the United States had a continuing interest in repayment.

The agreement to extend the treaty was signed on March 28, 1917 by Haitian Minister for Foreign Affairs Louis Borno (later president of the Republic) and American Minister to Port-au-Prince Arthur Bailly-Blanchard, who had returned to the post. This made the new expiration date May 3, 1936 — which would be 20 years from the formal exchange of ratifications. The extension was vigorously assailed by treaty opponents on two grounds: (1) only 11 months had passed since the treaty had come into legal effect, the conclusion that it would not in the original 10-year period of its life "fully accomplish" its purpose was not justified; and (2) the extension of the treaty was not presented for action to either the Haitian Congress or the U.S. Senate. Article XVI, however, provided for extension of the treaty under given circumstances and that provision had been accepted by the respected legislative bodies without dissent.

Though the treaty was duly extended, the hoped-for loan did not materialize. This further embittered the Haitian government toward the United States. When the new Haitian Congress assembled in April 1917, it was presented with the U.S.-approved draft of the new constitution. It appears that the Dartiguenave administration, in a momentary pique, let it be known that the proposals were actually U.S. demands. As a result, they were summarily rejected and the Haitian congress began work on a revised constitution, taking time off, however, to reject, unanimously, an American-inspired proposal for a declaration of war on Germany.

On June 16, President Dartiguenave suggested to American officials that the Haitian Congress be dismissed and the proposed constitution be submitted to the people in a referendum. News of this reached the newly elected legislators, sitting as a national assembly, and they rushed to complete their draft of a new constitution. The reading of their draft was in progress when Chief of Gendarmerie, Brig. Gen. Smedley Butler, entered the chamber and delivered to the speaker Dartiguenave's decree dissolving the assembly. When the speaker refused to accept it, Gen. Butler read it to the assembly himself in impeccable French. He then ordered the chamber cleared.

With the assembly dispersed, constitutional reform proceeded at a more leisurely pace. It was evident that provision would have to be made for the operation of governmental affairs in the absence of a legislative branch. Further, it was clear that no legislature could be elected that was inclined to cooperate with the Occupation. The task of drafting a new Haitian constitution thus fell, notwithstanding Franklin Roosevelt's later claim to authorship, to three American officials directly involved with the Occupation. They were Rear Adm. H.S. Knapp (the senior naval officer in Haiti and in Haitian waters), Brig. Gen. Eli Cole (the Marine Brigade commander), and F.L. Mayer (U.S. chargé d'affaires in Port-au-Prince). Roosevelt's role seems to have been limited to a cursory review of work done by the officials named above.[1]

The drafting proceeded into the fall of 1917 and included frequent consultation with Dartiguenave administration officials, particularly Louis Bono, minister for foreign affairs. The challenge was to produce a document, democratic and liberal in tone, that could still be used to validate unrepresentative government.

The constitution ultimately produced begins with the words, "The Republic of Haiti is one and indivisible, free, sovereign and independent. Its territory, including the islands dependent thereon, is inviolable, and cannot be alienated by any treaty or by any convention." It provided for a popularly chosen Senate and Chamber of Deputies, and for the election by these two bodies meeting in joint session as the national assembly of a Haitian president whose term of office would be four years (instead of seven years as specified in the previous constitution). Deputies were reduced in number from 72 to 36, and senators from 39 to 15. Provisions were made for a council of state composed of two

1 F. L. Mayer to Adm. H.S. Knapp, Aug. 12, 1918.

members appointed by the Haitian president. In addition to its regular duties of advising the president, it would exercise the legislative power in the absence of the elected legislature.

Freedom of the press, trial by jury and the right of assembly were guaranteed. The sensitive issue of foreign ownership of land was addressed in Article V, which reads: "The right to own real estate shall be given to foreigners living in Haiti and to the societies organized by foreigners for purposes of residence and agricultural, commercial, industrial or educational enterprises. This right shall cease after a period of five years from the date when the foreigner shall have ceased to reside in the country or the activities of said companies shall have ceased."

A special article ratified the acts of the American Occupation. The Gendarmerie d'Haiti was established as the country's only legal armed force.

The joker in the U.S.-drafted Haitian Constitution of 1918 lay in Title VIII, the "transitory provisions." After establishing the term of the current president so that it would expire on May 15, 1922, Title VIII provided that, "the first election of members of the legislative body after the adoption of the present Constitution shall take place on Jan. 10 of an even-numbered year. The year shall be fixed by a decree of the President of the Republic published at least three months before the meeting of the primary assemblies. The session of the legislative body then elected shall convene on the constitutional date immediately following the first election."

This, coupled with the provision giving the legislative power (including the power to elect the president) to the president's appointed council of state, actually made it possible for the incumbent president to remain in office indefinitely and to rule without benefit of an elected legislature by the simple device of not naming a date for legislative elections and by packing the council with his own men. A final provision under Title VIII suspended the non-removal of judges for a period of six months from the date of the promulgation of the new constitution.

These provisions, all of them, were carefully thought out. American experience with the last elected Haitian legislature made it clear that the options were establishment of open U.S. military government or support of an

American-influenced Haitian dictatorship. A third option, evident today with the advantage of hindsight, was withdrawal of U.S. forces. That, apparently, received no serious consideration.

The legality of adopting the new Constitution by public referendum worried U.S. officials, especially Adm. Knapp. This likely explains the transitory provision concerning the Haitian judiciary. American authorities were quite prepared to pack Haitian courts to ensure that U.S. actions in Haiti were not thwarted by legal technicalities. Adm. Knapp put his misgivings in a letter to the Secretary of the Navy in which he expressed the view that as the Haitian Constitution then in effect would remain lawful until amended, the method of amendment should, in legal strictness, conform to the old Constitution. This would not be the case, he wrote, if the Constitution were changed by popular vote.

He suggested that this legal issue might be resolved by relying on the theory that all powers in democratic government come originally from the people, and that if this were impossible by the regular method prescribed in the constitution to amend it, the matter could be left up to the people as a whole.[2]

Adm. Knapp's philosophical loophole, if one were needed in Washington at that time, was adopted, though the circumstances surrounding the constitutional referendum when it was actually held stripped it of what moral justification it might possibly have had. American influence in the adoption of the new Constitution, as in the signing and ratification of the treaty, poorly clothed the bald fact that ends were being taken to justify means.

On June 12, 1918, the referendum was held under the watchful eye of the Gendarmerie. As the overwhelming majority of those voting could not read or write, ballots did not have to be marked. White ballots, signifying approval, were handed to each voter as he approached the ballot box. Pink ballots, signifying disapproval, were available from the attending gendarmes upon request. Entertainment, food, drink and other persuasions were used to get out the vote. At some polling places, to discourage repeat voting, the thumbs of voters were marked with indelible ink as they left the polls.

2 Rear Adm. H.S. Knapp to the Secretary of the Navy, Sept. 8, 1917.

Col. John H. Russell, USMC.

Gendarmes had received strict instructions not to use influence to obtain affirmative votes, but officers were told they should explain the Constitution to the people who were unfamiliar with its provisions. Considering the large Haitian population and the fact that voters were not registered, the turnout of just fewer than 100,000 was small, even though it probably was the largest vote ever cast.

Marine Brigade Commander Col. John H. Russell reported that the Dartiguenave administration used its influence to obtain the adoption of the Constitution and "it is thought all government employees voted at least once for it."[3] It seems clear that most did not know what they were voting for. Russell reported that one man when questioned thought that he was voting for the election of a Pope. Another thought that he was voting for the president. It is likely that most of the Haitian elite class boycotted the referendum altogether.

Under the circumstances it is not surprising that the Haitian people overwhelmingly approved the new Constitution by 98,294 to 796. The Haitian Constitution proclaimed and adopted, Occupation authorities proceeded to carry out their obligations under the treaty. Soon, however, there was more trouble with the cacos.

3 Col. John H. Russell to the Chief of Naval Operations, June 17, 1918.

11

THE CACO WAR

Haitians, the day like the last of January 1804 will soon rise. Since four years the Occupation insults us in every way, every morning brings us a new sadness. The people are poor and the Occupation is pressing them down with taxes; it is spreading fires and preventing people to rebuild in wood under the false pretense to beautify the town. Haitians, let us be firm, let us follow the example of Belgium. No matter if our towns are burned. For it is not a vain thought that was written on the grave of the great Dessalines: "Upon the first shot the towns disappear and the nation rises." The holy cause which is spreading in the North has for chiefs men of good sense. The South is waiting for only one good man to follow this sublime example. No fear! We have arms! Let us drive out that ravenous people whose ravenousness is represented in the person of their President Wilson, traitor, vagabond, rioter, thief, you will die with your country.

Hooray for independence! Hurrah for union! Hurrah for legitimate war! Down with the Americans!

— Caco proclamation posted in Port-au-Prince, March 15, 1919.

The immediate cause of the resumption of hostilities with the cacos after almost two years of calm has been laid to certain abuses in the system of *corvee* inherited from the French colonial administration whereby the Gendarmerie d'Haiti undertook a program of widespread repair and construction of roads. Haitians were required to perform unpaid labor on the roads in their district, and in this manner some 630 miles were built or repaired at an average cost of $205 per mile.

Prior to the Occupation the corvee had not been consistently enforced for generations. In 1916, probably at the suggestion of the Haitian minister of public works, the system was revived. Brig. Gen. Smedley Butler described it as having a barn-raising atmosphere, with dances and games in the evenings, and prizes of little flags to hang on gateposts where the roadwork was particularly well done.[1] With President Dartiguenave, Gen. Butler inspected the work from an emblazoned Ford touring car. Dartiguenave gave talks of encouragement to the laborers. Men, women and children kept holes in the road filled with stone and gravel.

On Jan. 1, 1918, a caravan of 27 automobiles and trucks, headed by the Haitian president's car, made the 180-mile trip to Cap Haitian from Port-au-Prince in 13 hours, 50 minutes. The caravan stopped en route for receptions at six towns. This probably was the first time wheeled vehicles traveled the entire route. In French colonial days, approximately six miles of the route through swamps required rafts or horses.

Initially, there were few complaints about the new roads but as work spread into sparsely populated areas, laborers were required to work outside their home districts, and competition among supervisors led to men being pushed beyond their endurance. Some were required to work for hours exceeding the lawful limits. Enforcement was a made-to-order issue for opponents to the Occupation. Stories of excesses, often greatly exaggerated, spread with dark hints that the corvee was but the opening gambit of an American campaign to restore slavery.

Those who fled to avoid work on the roads joined the remnants of caco bands in the hills. Isolated incidents of terrorism were reported. On Oct. 1, 1918, the corvee was abolished, though either by a misinterpretation or disobedience of the order, it continued in the Maissade-Hinche District in the North until March of the following year.

A rare commodity in Haiti during the Occupation was native leadership. On Sept. 3, 1918, however, a convict persuaded his gendarme "chaser" at a work project to accompany him in an escape into the hills. The convict was Charlemagne Massena Peralte, a man of great charisma and spiritual intensity. He quickly rallied all the cacos in the North under his banner and vowed to drive the Americans into the sea.

1 Statement of Brig. Gen. Smedley D. Butler to Select Committee on Haiti and Santo Domingo, Oct. 31, 1921.

Col. Russell, the Marine Brigade commander, first reported encounters with Charlemagne's forces in October. Thirty-five cacos and two Haitian gendarmes were killed. In what was for Col. Russell a rare miscalculation, he wrote, "This affair has no political or military significance whatsoever, and repetition of such raids by bandits hidden away in the hills on the frontier may be expected for some time to come."[2] He estimated the strength of the guerrillas to be 2,000 poorly armed men whose weapons were mostly machetes, knives, pikes, a few pistols and up to 300 rifles. He said the enemy was short on ammunition, and their rifles were mostly German Krags to which they adapted captured or stolen ammo by wrapping goatskin on string around the cartridge bases. Col. Russell accordingly issued strict orders concerning the accounting of gendarme and Marine Corps ammunition.

Despite their great handicaps, Charlemagne's command continued to grow. His personal magnetism was such that even prior to the bizarre circumstances that attended his death, he was a messiah in the eyes of Haitian peasants. By the spring of 1919 his strength had grown to between 5,000 and 10,000. Later in the year it peaked at 15,000. At first he was opposed only by the Gendarmerie, but the growing seriousness of the revolt caused the Marine Brigade to be ordered into action in March 1919. Total brigade strength was only 1,000 officers and men. This meant that after garrison and logistics demands were met, only a few hundred were available for field operations. Further, few Gendarmerie units could be spared to carry the fight, for they were organized to function only as a local police force.

The Marine Brigade commander sent an urgent request to reinforce his command by at least a battalion of men. But Secretary of State Lansing, then at the Paris Peace Conference with President Wilson, considered that politically inexpedient. Haiti already had become something of an embarrassment to the United States on the world stage.

In early October, Charlemagne and a sizeable force under his command encamped 15 miles north of Port-au-Prince. On Oct. 7 a small band of cacos entered the capital and set fire to buildings in the business district near the

2 Col. J.H. Russell to Maj. Gen. Commandant, Oct. 17, 1918.

Sgt. Herman H. Hanneken, USMC.

foreign legations. The self-styled "Chief of Revolutionary Forces against Americans on Haitian Soil" sent an ultimatum delivered through the British Legation to the diplomatic community in Port-au-Prince. It demanded the surrender of the Haitian government to Charlemagne.

U.S. Marines quickly drove the cacos force back into the northern mountains. A reward of several thousand dollars was issued for Charlemagne's capture, and efforts to destroy him were redoubled. The story of his entrapment and death is one of the most thrilling in Haiti's history.

The commander of the Grande Riviere District in northern Haiti in August 1919 was Herman H. Hanneken, a Marine sergeant as well as a Gendarmerie captain. One other American served at the post as well. He was Marine Cpl. William R. Button, who also served as a lieutenant in the Gendarmerie. Early in the month, Hanneken made an arrangement to capture Charlemagne with the help of Haitian ex-Gen. Jean Conze, gendarme Jean Francois (known as the "Secretary") and Haitian Cherubin Blot.

The scheme was as follows:

The Haitians would establish a camp at Capois, near Grand Riviere and stage mock raids on police outposts in the vicinity. Having established a reputation as rebels, they would invite Charlemagne to participate in a large-scale assault on Grande Riviere. During that time Charlemagne was to be ambushed by gendarmes brought in from Cap Haitian.

As a result of their activity and numerous invitations, Conze and friends were visited by minor cacos chiefs in September and October. They entertained lavishly, Hanneken having provided the camp with $75 of *tafia* (native rum) and $10 worth of cola. These and other expenses, including blue cloth for caco uniforms, he paid for personally.

Early in October, the real cacos became suspicious of the camp at Capois, and to save the situation Hanneken made a simulated attack there during which he pretended to be wounded. This and a $200 contribution to the revolutionary cause had the desired effect. The money allegedly came from sympathizers in Grande Riviere, but Hanneken actually paid it. On Oct. 26, Charlemagne and his brother, St. Remy Peralte, with many other chiefs and about 1,200 fighters, arrived a Capois.

It was arranged that the attack on Grande Riviere would take place on the night of Oct. 31, and that Charlemagne would await the results at Mazare, about three miles outside the city. Due to the unexpected size of the attacking force, Grand Rivierie was secretly enforced with an additional Marine machine-gun crew. Early that evening Hanneken and Button, their faces blackened, wore ragged clothing and took up positions at Mazare with 20 handpicked gendarmes. Hanneken was armed with a 45-cal. automatic pistol and a 38-cal. revolver. Button carried a Browning automatic rifle. The others had carbines.

At about 10 p.m. some 700 cacos passed them en route to Grande Riviere. From their ranks dropped Jean Francois, the "Secretary." He told Hanneken that Charlemagne had changed his mind and instead of coming to Mazare, he would await news of the attack at his mountaintop camp, a hard three-hour march away. The Secretary had been given the countersign "General Jean" and knowledge that Charlemagne had arranged with the attack force that a detachment would be sent up the mountain to inform him when the city had fallen.

Hanneken decided that he and his men would pretend to be the returning detachment. With the Secretary in the lead, they started off, and after three hours of difficult mountain climbing they arrived at the first of Charlemagne's six outposts. They were halted there while the Secretary hurried on to inform Charlemagne that the detachment from the Grande Riviere victory had arrived. Charlemagne ordered that the detachment report directly to him, and the Secretary returned to tell Hanneken. He warned that it would be extremely dangerous to proceed in view of the numerous outposts through which they

Cpl. William R. Button, USMC.

would have to pass. Nevertheless, the detachment continued up the mountain with the Secretary in the lead followed by Hanneken, Button and the gendarmes bringing up the rear.

The first outpost was passed without incident, the countersign being given by each man as he hurried by. After walking about five minutes they were stopped at the second outpost and scrutinized closely by the caco in charge. Hanneken and Button acted as if they were exhausted and could not speak. Their act must have been good, for they heard comments about how tired they seemed to be.

They passed right through the third and fourth outposts, but the fifth was a problem. A man with a revolver halted them. Hanneken, still pretending to be exhausted, staggered through but Button was seized by the arm and asked where he had obtained such a nice rifle. Button jerked free, narrowly avoiding detection. Muttering some choice Haitian oaths, he hurried on after Hanneken. The last outpost was the immediate guard around the supreme chief of the cacos who stood next to a campfire about 30 paces distant. The Secretary pointed out Charlemagne and stepped aside. Hanneken and Button moved to within 15 feet of the guard and heard rifle bolts being cocked. Hanneken took deliberate aim at Charlemagne while Button cleared the camp using his automatic rifle. The gendarmes sprang into action, taking up positions to repel a counterattack.

Charlemagne fell with a bullet in his heart. Nine cacos were killed and an undetermined number of others were wounded. Nine serviceable rifles, three revolvers, 200 rounds of ammunition and seven swords were seized along with

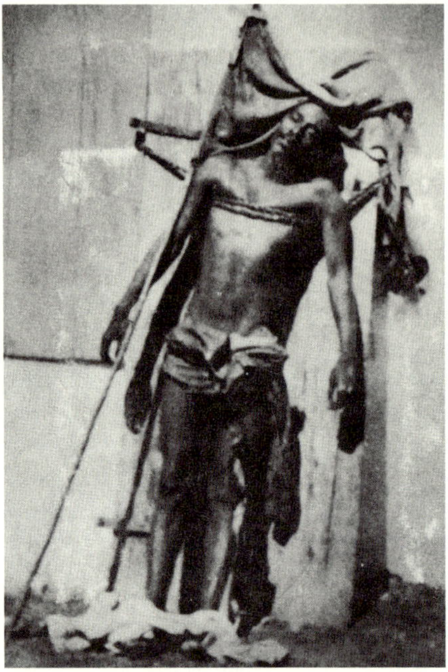
The body of Charlemagne.

much informative correspondence.[3] Hanneken and his men delivered Charlemagne's body to Grande Riviere at 9 a.m. the day after Halloween Eve.

At Grande Riviere the body of the caco chief was laid out on a door, arms outspread, and photographed. Prints were widely distributed by the Gendarmerie to prove the mystic leader, who claimed to be immune from Marine bullets, was dead. These photos triggered a widespread story that the Americans had crucified Charlemagne. They myth is still believed in Haiti. The photos eventually reached the United States, where they fueled growing opposition to the Occupation.

To thwart necromancy, Charlemagne's body was interred in a great block of concrete. After the Occupation ended, the concrete was broken up and his remains were given a state funeral. The ceremony became the subject of a famous painting by one of Haiti's renowned artists.

For their part in killing the caco chief, Hanneken and Button were awarded the Medal of Honor. Sgt. Hanneken was later commissioned in the Marine Corps and retired as a brigadier general.

Charlemagne's death virtually ended the war in the north, but his successor, Benoit Batraville, rallied the cacos in other parts of the country. In the early morning hours of Jan. 15, 1920, he led a large force into Port-au-Prince and burned more than a city block in Belaire, one of the poorest suburbs. The Marine garrison routed the cacos, killing approximately 300 of them in the

3 This account was drawn principally from action reports filed by Hanneken and his superiors in the Gendarmerie.

process. In weeks that followed, the movement's back was broken and more than 3,200 prisoners were taken.

Benoit evaded capture for months. In April when he and his band ambushed a patrol and captured its leader, Marine Sgt. Lawrence Muth, who was commissioned as a lieutenant in the Gendarmerie. A caco defector who witnessed his execution reported Muth's fate. Muth was wounded during the ambush but still conscious when examined by Benoit, who then nearly severed his head with a machete. Benoit then directed a Dominican chief named Francique to remove the sergeant's brain, which was smeared on the cacos' rifle sights with the claim that this would improve accuracy. Muth's heart and liver, believed to be seats of courage and sagacity, were removed and eaten. There were at least three confirmed reports of such cannibalism of Marines during the Caco War.

Efforts to track down Benoit were redoubled, and on May 20 a Marine patrol, acting on a reported sighting of the man, cornered and shot him in his mountain camp. This signaled the end of the war. The remaining chiefs quietly surrendered and peace was restored throughout the republic.

The Marines and the gendarmes defeated the cacos by adopting a plan of operations that broke up their field forces into 20- to 40-man patrols that methodically tracked down the enemy on their own terrain. It was the hardest sort of operation imaginable. Though greatly outnumbered throughout the two-year period of hostilities, these patrols had the decided advantages of better arms, better discipline and better training. They operated on the cardinal principle of never losing their weapons to the enemy. They never deliberately alienated the people living in battle areas, and in all of the atrocity stories that later circulated, none purports to describe acts of violence taken against a village, a peasant's shack or his crops.

In retrospect, the Caco War could be described as "clean." Seven Marines and 27 gendarmes were killed in action compared to an estimated 2,000 cacos. In late 1919, at the height of the war, the commandant of the Marine Corps asked that the Haiti Campaign Badge, previously authorized for service in Haiti from July 9 to Dec. 6, 1915, be awarded to Marines then engaged in operations against the cacos. On Nov. 6, 1919, U.S. Secretary of the Navy

Josephus Daniels replied as follows:

"It is not considered that the services rendered in Haiti since the limiting date in General Order 305, namely Dec. 15, 1915, have been sufficiently hazardous or the military importance necessary to justify the extension of that date." Haiti had all the markings of a Class A political issue, and the Wilson administration had ample reason to downplay military operations there.

12

OCCUPATION DIPLOMACY

The Haitian government was painfully surprised at hearing that, in the existing differences between it and the American minister to Haiti, the government of the United States only regretted the reason which compelled its representatives to arrive at the grave decision of stopping payment of the salaries of the president of the Republic, the secretaries of state and members of the Council of State.

And what is that reason? The non-cooperation of members of the Haitian government with American officials....

The Haitian government, much to its regret, can only consider the suspension of payment of the salaries of the president of the Republic and other members of the government as an unwarranted and coercive measure intended to compel it to adopt financial and other schemes which it regards as absolutely antagonistic to the well-considered interests of the Haitian people....

— President Dartiguenave to President Woodrow Wilson, Aug. 9, 1920.

The adoption of the new Haitian Constitution did not usher in a new era of cooperation between the Haitian government and Occupation officials. Public works, public health, public safety, the customs and to a large extent the administration of public finances were now firmly in American hands. But Occupation officials frequently found themselves frustrated by Haitian courts and the balking government, which as often as not refused to provide the enabling legislation necessary to carry out "treaty obligations."

Within a few days after the proclaiming of the Constitution, U.S. Minister Bailly-Blanchard called on the Haitian president and told him all future legislation bearing in any way on the objectives of the treaty must be cleared in advance by the U.S. Legation. Further, the legation reserved the right to introduce legislation deemed necessary for the proper observance of the treaty.

In October 1918, U.S. Financial Adviser Addison T. Ruan was incensed by persistent irregularities in accounts administered by the Haitian minister of finance. Ruan demanded that Haiti's Council of State enact a law forbidding the minister from approving vouchers that did not have Ruan's prior authorization. The council refused on the grounds that doing so would be in violation of the Constitution. The Marine Brigade commander thereupon ordered the National Bank of Haiti to stop all payments to the Haitian government.

This impasse continued for three months during which American officials refused to pay the funeral expenses of the Haitian minister in Washington who had died there. This unpleasantness was mitigated in part but only for a short time, when Ruan resigned and his successor, John H. McIlhenny, arrived on the island. Relationships creaked along on an ad hoc basis while American patience with an evasive and uncooperative President Dartiguenave wore increasingly thin.

In August 1920, after discussions with Washington, McIlhenny abruptly suspended salary payments of senior Haitian officials, including the president. Dartiguenave appealed directly to President Wilson and received in reply a lecture delivered by Bailly-Blanchard. The American Minister to Haiti told him that Wilson regretted that the close coöperation that previously existed between the two governments had recently and unfortunately been lacking. Bailly-Blanchard emphasized the importance of strict adherence to the treaty agreements. The lecture ended with an offer to resume salary payments provided the Haitian government repeal 11 laws and enact four new ones.

The ones the Occupation wanted repealed dealt with such matters as ownership of real estate by resident foreigners, the return of sequestered property to Germans, pensions, motor vehicle duties, trademarks, mines and mining concessions, preparation of primary school teachers, manual training schools, railroads and trams, and firearms and ammunition. The new laws that were demanded concerned currency reform and the lease of state-owned land.

The Haitian authorities ultimately agreed to the demands and U.S. control of legislation was assured. Various financial and commercial arrangements soon followed.

National City Bank of New York took total control of the National Bank of Haiti while buying out French interests for $1.4 million. A commission was named to settle foreign claims against Haiti. Also, a $40 million bond issue, to be serviced by a first claim on internal revenue and a second claim on customs receipts, was approved. But the bond issue was to take effect only after market conditions improved.

In addition, a concession for oil exploration was granted to the Sinclair Corporation. However, Occupation authorities rejected a proposed American company's timber contract because it had monopolistic features. In all of these matters great pains were taken to protect the interests of Haiti, and there is virtually no evidence to support claims that Americans used their positions with the Occupation for personal enrichment. The claims were freely made but ill-substantiated by Haitian and American critics of the Occupation.

In April 1921, Edwin Denby, the new Secretary of the Navy, toured Haiti and was, like Franklin D. Roosevelt before him, favorably impressed by signs of progress. At the town of Maissade he was surprised to find a street decorated in his honor with flowers and palms and a triumphal arch saying *Vive l'Occupation!*

A delegation of local officials, teachers and other citizens turned out to give Denby a cordial welcome. Children sang the Haitian anthem as well as one verse in English of "America the Beautiful." A flower bouquet was presented to the secretary along with a written request to continue the Occupation. Similar ceremonies of various degrees of formality greeted him in other towns. [1]

It should be noted, however, that the triumphal arches, flowers, songs, speeches and petitions echoed those prepared in the past for a Dessalines, a Faustin or a Vilbrun Guillaume Sam.

1 Secretary of the Navy to Secretary of State, April 6, 1921.

13

THE PURLOINED LETTER

From enthusiasm to imposture the step is perilous and slippery; the demon of Socrates affords a memorable instance how a wise man may deceive himself; how a good man may deceive others; how the conscience may slumber in a mixed and middle state between self-illusion and voluntary fraud.

— Edward Gibbon, "The Decline and Fall of the Roman Empire."

The Commandant of the Marine Corps in 1919 was Major Gen. George Barnett, an extremely hard-working and conscientious officer. In September of that year while reviewing the records of courts martial convened by the First Provisional Brigade in Haiti, he was struck by certain allegations made by the defense counsel, Lt. F.L. Spears, in the trial of Privates Walter E. Johnson and John J. McQuilkin Jr.

The men were charged and convicted of having struck Haitian prisoners in their custody. Their defense was that they had done so in obedience to the orders of their superior officer, one Sgt. Brokaw, USMC. Brokaw had later shot and killed the prisoners, but was not tried due to reasons of insanity, for which he was then confined in a mental institution.

The allegations by Lt. Spears, who had been in Haiti but a few months and in the Marine Corps less than a year, were that indiscriminate killings of Haitians had occurred over a period of years. Gen. Barnett accepted this at face value because, as he later explained, "In all my experience in 44 years in the service I have taken it as a matter of course that a statement made to me by a commissioned officer was true."[1]

The general immediately drafted a letter to the Brigade commander in

1 Statement of Maj. Gen. George Barnett, USMC, to Select Committee on Haiti and Santo Domingo, Oct. 24, 1921.

Port-au-Prince, Col. John H. Russell, who had recently returned to Haiti at the request of Secretary of State Lansing for a second tour.

By virtue of his prior experience he was considered the most knowledgeable officer on Haitian affairs then in the service. What's more, he was known to have the confidence of the Haitian government. Gen. Barnett's letter ordered Col. Russell to take immediate steps to end "unlawful executions of Haitians," and to make a thorough investigation of the charges made by Lt. Spears in his argument as defense counsel in the Johnson-McQuillin courts martial. This was followed by a second letter marked "personal and confidential" on Oct. 2. Parts of this letter are produced below:

"My dear Colonel:

"Since you left here several things have come to my notice with reference to the affairs in Haiti, especially in relation to the duties of the gendarmes in the interior. The court-martial of one private for the killing of a native prisoner who brought out a statement by his counsel — which showed me that practically indiscriminate killing of natives has gone on for some time....

"I was shocked beyond expression to hear of such things and to know that it was at all possible that duty could be so badly performed by Marines of my class....

"I want personal instructions sent to every officer and non-commissioned officer, both Marines and Gendarmerie, that conditions as shown by the evidence in the trial of the above private referred to must be corrected, and such action cannot be tolerated for a moment; and I want every case thoroughly sifted and the guilty parties brought to justice. I think this is the most startling thing of its kind that has ever taken place in the Marine Corps, and I don't want anything of the kind to happen again. I think, judging by the knowledge gained only from the cases that have been brought before me, that the Marine Corps has been sadly lacking in right and justice, and I look to you to see that this is corrected and corrected at once."[2]

Though the letter had been marked personal and confidential, Gen. Barnett had a copy of it placed in the official files at Marine Corps headquarters in

2 Maj. Gen. George Barnett to Col. John H. Russell, Oct. 2, 1919.

Maj. Gen. George Barnett, USMC.

Washington, along with a copy of his earlier official letter.

Col. Russell acknowledged receipt of both letters and began a thorough, but due to certain inherent difficulties in such matters, time consuming investigation. Sworn statements had to be secured from personnel no longer serving in Haiti, transportation arranged, etc. As a consequence, when Gen. Barnett was relieved of duty on June 30, 1920, Russell's findings had not yet been received in Washington.

By then, the performance of U.S. Marines in Haiti, and the Wilson administration's responsibilities associated therewith, had become a front page, political issue in the United States. Wilsonian idealism, of which the country was growing tired as his second term neared an end, was dragged through the mire of Haiti. The most lurid tales — and most baseless charges — circulated as gospel. People seemed prepared to believe anything, provided it was damning to the Haitian Occupation or to America's armed forces. But then, even Gen. Barnett, with his 44 years of service, had been quick to accept the worst.

The summit of irresponsible journalism likely was reached when Herbert J. Seligman wrote an article published in *The Nation* on July 10, 1920. It was titled "The Conquest of Haiti." Seligman had made a brief visit to Haiti earlier in the year and returned to the United States nearly rabid on the subject of American Marines and their role in the Occupation.

"The history of the American invasion of Haiti," he wrote, "is only additional evidence that the United States is among those powers in whose international dealings democracy and freedom are mere words, and human lives negligible in face of racial snobbery, political chicane and money. The five years of American Occupation, from 1915 to 1920, have served as a commentary upon the white civilization which still burns black men and women at the stake. For Haitian men, women and children, to a number estimated at 3,000, innocent for the most part of any offense, have been shot down by American machine guns and rifle bullets; black men and women have been put to torture to make them give information; theft, arson and murder have been committed almost with impunity upon the persons and property of Haitians by white men wearing the uniforms of the United States.... In this five years' massacre of Haitians less than 20 Americans have been killed or wounded in action....

"The Haitians in whose service United States Marines are presumably restoring peace and order in Haiti are nicknamed 'Gooks' and have been treated with every variety of contempt, insult and brutality....

"This militarist, imperialist burlesque of the profession with which the United States entered the war on behalf of weaker states leaves the Haitians little to do but wonder what the United States intends.... In the absence of any plans for Haiti's regeneration except through 'development' of the country by exploiters, the Haitian may derive what spiritual nourishment he can from the Wilsonian phrases with which the United States thuggery disguises its deeds."

"The Conquest of Haiti" created a sensation. The State and Navy departments were flooded with demands for information. The National Association for the Advancement of Colored People denounced the administration's policy in Haiti. It being an election year, and Sen. Harding knowing an issue when he saw one, a chorus of Republican demands for investigation of the "Haiti Scandal" soon flooded the nation. Gen. Barnett was recalled to Washington from his home at Huntley, Va. He arrived on the night train early on the morning of Sept. 18, 1920.

In obedience to orders from Secretary of the Navy Josephus Daniels, he spent the day compiling a report on Marine activities in Haiti during his tenure there. Documenting his report were copies of his correspondence with Col. Russell, including the personal and confidential letter of Oct. 2, 1919. The complete report was delivered to Secretary Daniels that evening, in the pres-

Secretary of the Navy Josephus Daniels.

ence of Daniels' public relations assistant, a Mr. Jenkins.

It was Jenkins' responsibility to read the report and help prepare Daniels to field questions. What he thought when he stumbled across Gen. Barnett's personal letter to Col. Russell and its "practically indiscriminate killing of natives" part is not known, but on the basis of circumstantial evidence, opportunity and philosophical inclination, he is a prime suspect in the mystery of who leaked it to *The New York Times*.

When the story broke in the middle of October just a few weeks before the election, Gen. Barnett was en route to the West Coast. Daniels' summons reached him in Chicago and he caught the next train back east. The meeting between the Secretary of the Navy and the retired commandant was a memorable one.

"Mr. Daniels, I see in the morning papers that you have stated that you never saw my letter until it was published."

"I never saw it."

"Excuse me, sir, but you did see it."

"I never saw it."

"You did."

"Of course, General, if you say I saw it I must have seen it, but I forgot it."

"You did not say that you had forgotten it, but that you had never seen it."[3]

3 Statement of Maj. Gen. George Barnett, USMC, to Select Committee on Haiti and Santo Domingo, Oct. 24, 1921.

A naval court of inquiry, headed by Rear Adm. Henry T. Mayo, was immediately appointed to investigate the performance of Marines in Haiti. It met in Washington in October and journeyed to Haiti shortly thereafter, arriving in Port-au-Prince on Nov. 8 aboard the USS *Niagara*.

Republicans and editorialists in *The Nation* charged that the issue was whitewashed before the first witnesses appeared before the court. The findings of the inquiry when they were released, after the Republican presidential victory, were interpreted as nearly complete vindication of American military conduct in Haiti. The court found that two unjustifiable homicides and 16 other serious acts of violence against citizens of Haiti had occurred during the period 1915-1920, and that in every case those responsible had been brought to trial before a general court-martial, convicted and sentenced. It further found that there were no proper grounds for the statement that "practically indiscriminate killing of natives has been going on for sometime."

Gen. Barnett took this as a rebuke to himself. He later petitioned Secretary of the Navy Daniels to state publicly that he had not been reprimanded. Daniels, however, was terrible in his self-righteousness. There is no record that he ever again addressed himself to Gen. Barnett on this or any other matter.

14

THE MISTREATED MISSIONARY

While Haiti's fertile soil now needs, keen human skill to till it;
What Haiti needs above all things, is Christian love to fill it.
God give thee health, and strength and grace, and favor with the Nation;
And never hide thee from His Face, but show thee His salvation.
Go brother, tell them of the Light, and of the Life Eternal;
That's offered unto black and white, in the bright world supernal.
We know, God shall be glorified whatever the condition;
With such a servant, true and tried, out in the Haiti Mission.
And if you'll not return to tell the thrilling Haiti Story,
May God "who doeth all things well" grant us to meet in Glory.

— Cyonfardd, "Prospectus for Proposed Evangelization of the Over Six Million Christless Negroes of Haiti and Liberia."

On Oct. 25, 1920, the Republican National Committee retained a lawyer to assist a Baptist missionary to "prevent any further suppression of the facts by Secretary Daniels or any other board which the present administration may create in its efforts to meet the charges of Senator Harding."[1]

It is doubtful if anyone was a greater thorn in the side of the American Occupation in Haiti during the last dozen years or so of its existence than the Rev. L. Ton Evans, a missionary of the Welsh Baptist Church. His letters and newspaper accounts of public appearances paint a picture of a man who, in the later stages of his life, was irrational in the extreme, a fanatic who became almost incoherent when he spoke or wrote on the subject of his Haitian "ordeal"

1 *The New York Herald,* Oct. 25, 1920.

at the hands of Americans in the Occupation. Yet for all of this he managed to fill a file of correspondence and reports at the U.S. Navy Department and a second one at the U.S. State Department.

In 1931 a bill, S. 2359, "For the Relief of L. Ton Evans, General Missionary and family, Haiti Mission, and proposed American Christian Industrial Tuskegee" was considered by the U.S. Congress. Had it passed, he would have received $300,000.

Throughout his long life, he wrote and talked to anyone and everyone who would read and listen to stories about his experiences in Haiti. A public figure who made the mistake of replying to his shotgun correspondence would thereafter be considered as not only a close personal friend of the pastor but also someone irrevocably aligned with him on every conceivable issue. British Pime Minister David Lloyd George, Theodore Roosevelt, several U.S. secretaries of state, and numerous U.S. senators and representatives found themselves in that embarrassing position, often for no other reason than that a courtesy had been paid by acknowledging receipt of one of Evans' letters.

He was born in Wales in 1863 the son of a wealthy saloonkeeper. After graduating from the Haverford-West Baptist Seminary in 1887, he became the organizing secretary of the District Gospel Temperance Council in his hometown. What his father, who by then had prospered to the extent of owning a brewery, thought of his son's chosen vocation is not known, but might well be imagined.

After several short-lived assignments in Wales during which he made a name for himself as a booze fighter, his church sponsored a trip for him to the Holy Land. He returned to accept in 1892 the position of general missionary at the Jamaica Missionary Board in Haiti, and for two years he performed missionary work in the Black Republic while based at Jacmel.

From 1894 to 1902 he served again in Wales, engaged as before in temperance activity. He then received a call from the First Baptist Church in Edwardsville, Pa. While there he filed papers to become a U.S. citizen "believing in a Republican form of government."

His aversion to alcohol almost provoked armed conflict in the small town of Edwardsville. At one point his church was set on fire, presumably by a pro-whiskey advocate. In 1908 he resigned his charge to become field secretary of the Baptist Mission in Haiti, spending four years in that post before returning to another small church in Pennsylvania as its pastor. During this period he became an avid letter writer on the subject of Haiti, and consistently spoke out in favor of American intervention "in the name of humanity" to end the senseless bloodshed he had witnessed while there.

He returned to Haiti for a third and last time on Thanksgiving Day, Nov. 29, 1917, as the superintendent of the Lott Carey Convention, a Baptist missionary group affiliated with and supported by an association of southern Negro churches. On arriving in Haiti, Evans was immediately struck by the great changes wrought by the Occupation. As he later testified before the U.S. Select Committee on Haiti and Santo Domingo, these changes went beyond roads and sanitation projects. The restoration of order, he believed, had transformed the life of the people. Where once there had been terror and bolted doors and shutters, he said, there was now peace and quiet, and pleasant, placid faces.

About six months after this return to Haiti, Evans made an inspection trip to certain churches in the north of Haiti, and one afternoon, as he was leaving St. Raphael with two Haitian churchmen, he encountered two white men on horseback who appeared to be having difficulty maintaining their seats. One of the horsemen was identified as the Gendarmerie chief in St. Raphael.

The renowned temperance fighter was mortified that a fellow representative of the United States should be seen in public in such a disgraceful condition. He said as much to the Brigade commander, Col. Russell, several weeks later during the course of an office call. His disclosure brought three quick responses: (1) A reprimand of the Gendarmerie officer, Second Lt. Askel H. Haug, also a U.S. Marine; (2) An order directing that the sale of alcoholic beverages to Marines in Haiti be forbidden; and (3) a complaint filed in the Haitian Court of First Instance at Cap Haitian by Lt. Haug charging the Rev. L. Ton Evans with "slanderous denunciation."

On July 13, 1918, Evans wrote to the Haitian judge, who had issued a summons to appear in court to answer the charge.

"Judge P.J.M. Obas, Juge d'Instruction au Tribunal Civil, etc.

"Dear Sir:

"I received your 'mandate' to appear at your court on the 29th instant, to be interviewed and questioned about somebody, etc.

"I shall be very glad if you will be so kind as to inform me definitely as to the name of the person and the exact nature of the complaint or the alleged charge to have been made in reference to myself which is a United States citizen and also in my position as General Secretary and Superintendent of the Baptist Mission work in Haiti and representing some 3 million negro Baptists and practically some 6 million more white Baptists of America as I do — I am entitled to this definite information from you and your court.

"I must say, however, that I have already planned to leave with my wife if possible by a Panama steamer leaving Port-au-Prince by the end of this month on official business to see my Missionary Board in the States as well as to visit our President, Woodrow Wilson, at Washington, it will be impossible for me to visit the Cape at the time you mention.

"I should also add that I have planned for a conference at Port-au-Prince with the Minister and our United States Superintendent of Instruction, who will arrange with Colonel Russell of the Occupation and probably with the President of Haiti, if not, National Council Remy, explaining to them the project of a proposed 'National and Industrial College,' the petition for which I took to the United States seven years ago, as, perhaps, the Judge knows.

"After the definite information asked, I may arrange what steps and when, I shall take on the return which I expect will be in October of this year."

The letter was signed, "Yours very sincerely, L. Ton Evans (For Christ and Haiti)."

The busy missionary then went to the United States, and apparently forgot all about his summons to appear in court.

On Dec. 28, 1918, Evans having returned from his trip, the Haitian wheels of justice took another turn. The Court of First Instance at St. Marc, where Evans then resided, issued a warrant for his arrest. What happened next is described in the missionary's own words in a letter to the secretary of the U.S. Navy, dated Nov. 29, 1922:

"1. This arrest took place Saturday afternoon, of Dec. 28, 1918 — by a ruthless invasion of Missionary's private study, at the home of Madame Orius

Paultre (widow of the late President Judge) and family at St. Marc, by the armed Corporal (native) of St. Marc, Gendarmerie — at the command of his white Captain Fitzgerald Brown, and while the Missionary Superintendent was peacefully pursuing his sacred Mission, and planning for native Churches and their Evangelical Schools, in Gonave and other places in Haiti, and while being in the very act of typing addresses on envelopes to be distributed to members of Churches, and branch mission stations the following Sabbath morning.

"2. The mere request of Missionary, to the armed native Corporal as to his Authority, and Right of forced intrusion upstairs to his private Study, and for production of his warrant showing cause of his arrest, was subsequently at the Gendarmerie headquarters construed by the excited American white Captain, under influence of liquor, into a Rebellion against Public Authority.

"3. Following the negro armed Corporal, who was also excited and somewhat savage — through liquor — the Missionary descended the stairway stepping to the open street, to find himself instantly surrounded by a heavy armed guard, of about a dozen Gendarmes. The little Negro Corporal roughly snatched the shade (parasol) from the hands of the missionary superintendent — as he opened it to protect himself under a broiling tropical sun — amid the cries of SHAME from the other armed natives who said 'Let the Missionary have his parasol.' He was immediately hurried through the public streets, past his Church, exposed to the scorching sun before the terror stricken natives, who fled to hide in fear, and trembling; many of whom were sobbing aloud. His further Request on passing his Church and in front of the door of the house of one of his native Deacons (and by this being convinced he was being led to prison, if not to be shot) to hand over the key to his locked study, so as to return same to native Dr. Hector Paultre (Clerk of local Church and Gonave mission) for the distribution of his addressed envelopes on Sabbath morning (next day) was not only savagely denied but intentionally DISTORTED (as declared by Haiti's Superior Court) at the same Gendarmerie headquarters, and by the same white, excited, cursing, and liquor-soaked American Captain, and United States Marine officer into an ATTEMPT of the Missionary TO ESCAPE FROM JUSTICE.

"4. At arrival at headquarters of St. Marc's Gendarmerie, Captain Brown, who had hurriedly drawn up the so-called two CHARGES, excitedly and rapidly read and hurled them (amid oaths, and threats of being shot, repeatedly aiming to lay hold of his loaded rifle) at the American Missionary as he stood in front of the Captain, and surrounded by the armed negro Gendarmes. After further insults, threats and cursing him for his efforts to Christianize, educate and industrially and morally develop 'the low down, dirty and damned niggers of Haiti' this United States Marine Officer assured the American Missionary — they (Captain Brown, General Alexander Williams, Major Welles, Lieutenant Haug, and other conspirators, it was understood) would not only discredit, degrade and absolutely crush him, and after an additional volley of oaths, and while raving like a drunken maniac, all the while laying hold of gun declaring I should be shot — he finally yelled 'Away with the villain, to prison.' The Missionary was dumbfounded at such a treatment, and sad humiliating spectacle — in striking contrast with that of Captain Brown's immediate predecessor — Captain Kinney, who himself had not only personally complimented the Missionary on his excellent work, and splendid influence among the natives, and Gendarmes — but as an expression of gratitude personally and voluntarily gave a special Official permit for him, and his trained Missionary wife, to conduct Sunday Religious services to prisoners at St. Marc old prison.

"5. No sooner than the prison was reached, than the American Missionary was searched, every article such as Bible, glasses, photographs of wife and little boys, photograph of United States PRESIDENT and State Department, etc., were taken from him after which he was thrust — that same Saturday afternoon — into a small, narrow, negro, slave, death-cell (of old French Colonial days) the heavy bar drawn with a tremendous thud, so as to strike terror no doubt into the poor superintendent Missionary's heart.

"6. Here, now he remained cut off from Family, Native Christians and the whole world — for 13 days, and 12 longest and darkest nights of his life, with armed negro pacing constantly before the small thick door of his dark dungeon, lying on hard bare floor of the old death-cell, hat in hand — panting for a breath of air; and amid the heart-breaking groans and yells of something like 180 negro prisoners (including the females, mothers and even babes), many of whom being beaten and tortured to death; then would follow scuffing, marching and heavy trampling as if wounded and dead were hurriedly and secretly carried away in the dark; and all the while the Missionary startled as the flash of the armed guard's small lantern was turned upon the narrow and tiny aperture (called window) near top of his cell door, and he heard the tramp of his feet — believing that the American Captain, and United States Marine officer in one of his drunken orgies, while entertaining his reputed drunken superior officers (General Williams, Major Welles, etc.) had at last given the stern command to drag the Christian Missionary from his dungeon before a firing squad as done with other prisoners in Saint Marc, and other Haitian prisons at this time without any pretense of trial...."[2]

The above tale, embellished and re-embellished, with some details added and others dropped, became the Rev. L. Ton Evans' standard sermon, and no doubt the poor old man believed the horror he had conjured up for himself. Certainly, many of his readers and listeners believed it. The records of the St. Marc Gendarmerie show that he was in fact confined in civil prison there from Dec. 28, 1918 to Jan. 9, 1919, and that he was provided with a mattress, a pillow and a blanket — luxuries not enjoyed by other prisoners or required to be furnished under Haitian law.

While in prison his sponsor, the North Carolina Mutual and Provident Association of Durham, discontinued its support of his missionary work, citing his political activity and threatening letters he had written to President Wilson. Among other things, the association wrote, "We want you to understand that the American Negro stands loyally behind the President and his action in shaping the destiny of Haiti, which is under our protection, and it places us in a very embarrassing position as a convention to be called in question as to our loyalty to the nation in a time like this.... It is not in the minds of the American negroes to undertake to shape the political policy of an independent country of which they know very little, especially in opposition to the American nation."

At his hearing before the Court of First Instance at Cap Haitian on Jan. 20, 1919, Evans recounted his story with the aid of an interpreter. After many years of missionary work in Haiti, he could speak neither French nor the native

2 Correspondence File of First Provisional Brigade in Haiti.

Creole. Lt. Haug, perhaps influenced by his superiors, dropped the charges and Evans was released from custody. He left Haiti shortly thereafter and never returned.

The story he later told — that he had been awarded a settlement in the amount of $100,000 by the Haitian Supreme Court for false imprisonment — has no basis in fact, though he enlarged and pressed that claim for years and, as previously stated, it eventually made its way to the floor of the U.S. Congress.

He was violently anti-Catholic, and in speeches and printed material he often compared the influence of the French clergy in Haiti to that of Rasputin in Czarist Russia. His prejudices were so open and so frequently expressed that it is difficult to see how his testimony on anything could be accepted at face value, and yet on Haitian matters he became a resident expert for those who attacked the Occupation.

The Rev. Evans never realized his dream of establishing in Haiti an "American Christian Industrial Tuskegee." His obsessive hatred of the Marines of the Occupation probably contributed to his physical and mental collapse, which forced his early retirement from missionary work. His last years were spent in Wyoming, Pa., where he died a bitter and broken old man whose longed-for martyrdom had somehow managed to elude him in that narrow prison cell at St. Marc in Haiti.

15

THE SENATE SELECT COMMITTEE

My dear Sir:

Some days ago I overheard the following conversation:

A. What do you think of the Hayti business?

B. Oh, I don't know.

A. Well, I will tell you this. A few years ago on a trip East, I met in the Broad Street Station in Philadelphia, an old friend of mine, an officer in the Marines. In reply to the question, as to what he was doing there, he said: "I am about to entrain a battalion of Marines for Hayti. There is some trouble down there, and official orders are to go down there and straighten things out. Unofficial orders are to kill every G.. D... nigger that comes in sight."

This may or may not have value to you. I send it as it may be of some use.

The remarkable letter quoted above, which might be described as hearsay thrice removed, was forwarded to the U.S. Secretary of the Navy by Sen. Medill McCormick on Oct. 21, 1921, with a terse, "I think this is of enough moment for you to read it through." Sen. McCormick had received it from James Weldon Johnson, secretary of the National Association for the Advancement of Colored People, who in turn had received it from a Mr. James Harold Coleman of New York, not otherwise identified.

James Weldon Johnson had been meddling in Haitian politics for more than a year. In April 1920, he visited Port-au-Prince with Herbert J. Seligman — an associate in the NAACP, a socialist and a contributing editor with *The Nation* — whose article "The Conquest of Haiti" has already been noted.

In Port-au-Prince, Johnson and Seligman closely associated themselves with the political opposition to the Dartiguenave government, especially with a

Sen. Medill McCormick.

newly organized group called the Union Patriotique, which was headed by George Sylvain, H. Pauleus Sannon, Sténio Vincent and Percival Thoby. This group claimed to represent virtually all of the Haitian people, but it is doubtful if it ever counted more than 10,000 adherents. Its leaders were articulate and literate men, and it is not surprising that they found in the two visiting Americans kindred spirits.

On their return to the United States, Johnson and Seligman persuaded friends at *The Nation* – Oswald Garrison Villard and Ernest H. Greuning – to aid them in welding together the resources of the publication, the NAACP and the Union Patriotique into the Haiti-Santo Domingo Independence Society, whose purpose would be to lobby and agitate for the immediate withdrawal of the Marines, the re-establishment of constitutional government and the negotiation of equitable treaties with the two republics.

Moral and limited financial support began to flow to the Union Patriotique from the Independence Society in the United States, and President Dartiguenave, believing his political opposition to be subversive in character, complained. There was little, however, that Occupation authorities could do, given the political atmosphere in the United States.

In December 1920 and again in February 1921, demands were made in the U.S. Congress for an investigation of affairs in Haiti. The U.S. elections over, it seems likely that a grumbling acceptance of in-house Navy Department inquiries would suffice (Adm. Knapp and Marine Gen. John Lejeune had filed reports separate from that of the Mayo court), had it not been for the arrival in Washington, in May 1921, of a Haitian delegation representing the Union Patriotique. Their avowed purpose was to present to the American government a "Memoire" reviewing the history of the Occupation as seen through Haitian eyes. It included certain demands the United States was to accept forthwith. In tone, the memoir might have leapt from the Rev. L. Ton Evans' typewriter with editorial assistance from *The Nation*. A few lines will suffice to show the level on which it was composed:

"... The American officers of the Gendarmerie are privates (in the American Marine Corps) who have been made officers in Haiti, and who have nothing but a most elementary education, which naturally renders them incapable of any military training.... The ghastly mortality in the prisons together with confirmation by survivors reveals a record of atrocities, of brutality and cruelty which defies description. It is a record for which it would be difficult to find a parallel.... It is the most terrible regime of military autocracy which has ever been carried out in the name of the great American democracy.... The Haitian people, during these past five years, have passed through such sacrifices, tortures, destructions, humiliations and misery as have never before been known in the course of its unhappy history."[1]

The Union Patriotique delegates were refused a formal hearing by the State Department, the proper and diplomatic thing to do, the Haitian government already being duly represented in Washington. This refusal brought strong protests from predictable sources. The complete text of the "Memoire" was

1 Union Patriotique, "Memoire," *The Nation*, May 25, 1921.

published, and the lead editorial in the same issue asserted that, "No impartial reader of the Haitian Memoire will doubt the truth of its charges." It warned that Haiti must not become either "America's Belgium" or "America's Ireland." Echoes of this sentiment converged on Washington from church federations, black groups and various pacifist organizations.

On Aug. 5, 1921, pursuant to a resolution of the U.S. Senate, a Select Committee on Haiti and Santo Domingo began hearings in Washington. The committee was chaired by Sen. McCormick (R-Illinois). Initially its members were Sen. Philander C. Knox (R-Pennsylvania), Sen. Tasker Oddie (R-Nevada), Sen. Atlee Pomerene (D- Ohio) and Sen. William H. King (D-Utah).

Before the committee appeared virtually all the leading figures of the intervention and Occupation. Adm. Caperton came — his mind not as sharp as it once had been — repeatedly having to refer to his voluminous papers and records. Roger L. Farnham — rich, successful, recently made a senior vice president of the National City Bank — explained the tangled web of Haiti's pre-Occupation finances. The Rev. L. Ton Evans — eyes flashing, voice quavering — read into the record his Haitian sermons, exasperating Chairman McCormick on more than one occasion. Gen. Smedley Butler regaled the committee with his little bits of Haitiana, disguising the steely side of his character that had earned him the nickname of "Old Gimlet Eye" in the Marine Corps. Gen. Barnett tried and failed to look something other than a fool over the "practically indiscriminate killing" remark he once accepted as gospel.

Hundreds of pages of testimony were taken. Those who appeared before the committee were subjected to sharp questioning not only by members of the committee but, remarkably, by a lawyer, Ernest Angell, representing the Haiti-Santo Domingo Independence Society, the NAACP and the Union Patriotique.

On Nov. 16, hearings adjourned in Washington and the committee prepared to resume its inquiries in Port-au-Prince. When the senators arrived there on Nov. 29, they were greeted by carefully staged demonstrations. Banners and posters were displayed asking in English, "Shall Haiti be your Belgium?" and "Shall Haiti be your Ireland?" The demonstrations almost certainly were a tactical error on the part of the Union Patriotique, which took credit for them. Sen. McCormick, for one, was openly angered.

Hearings were held in Port-au-Prince and in other Haitian cities through the first weeks in December, after which the committee traveled on to Santo Domingo. Sen. McCormick, who spoke fluent French, often dispensed with the services of an interpreter at sessions he chaired, and found occasion to criticize the almost total absence of French-speaking officers in Occupation forces.

The case against the Occupation presented by the Union Patriotique was weak. A great deal of the committee's time was taken up by witnesses who gave obviously coached and frequently hearsay testimony on such subjects as stolen mules, drunken Marines and a decline in public morals. Much of this backfired. Dr. George Sylvain, president of the Union Patriotique, spoke of the Occupation's failure to support Haiti's educational system. He stated with some degree of pride that Haiti had begun compulsory education in 1864. Further questioned, however, he admitted that only two percent of Haitians could read and write.

Also in Haiti at that time was Dr. Ernest Gruening. He called on Col. Russell, who noted in the Brigade Commander's Daily Diary Report that, "Mr. Gruening, I am afraid, has come to Haiti with the idea of not seeing anything good, but only looking for something sensational to write up for his paper."[2]

From informers, Col. Russell learned that during his stay in Port-au-Prince, Greuning made anti-American and incendiary speeches at the homes of Occupation critics. When he returned to the United States, Greuning put together a documented narrative titled "Conquest of Haiti and Santo Domingo," which was published in *The New York Times.* The article followed the lines of Herbert Seligman's earlier work and proved to be equally sensational.

When the committee returned to Washington, in the middle of December, a preliminary report was released in the form of this brief statement to the press:

"The members of the Committee are unanimous in the belief that the presence of a small American force in Haiti is necessary to the peace and development of the country as are the services to the Haitian Government of the American officials appointed under the treaty of 1915. There can be no abrogation of the treaty, and at this time any diminution of the small force

2 Daily Diary Report, First Provisional Brigade of Marines in Haiti, Nov. 10, 1921.

of Marines.

"It is important that steps should be taken forthwith to coordinate the labors of the representatives of the U.S. government in Haiti, and of the so-called American treaty officials. There should be appointed a Special Representative of the President, a High Commissioner, in whom should be vested the usual diplomatic powers of an envoy-extraordinary, and to whom, furthermore, all the American officials appointed under the treaty, as well as the commandant of the Marine Brigade, should look for direction and guidance. The members of the Committee know of no reason why the duties of the Financial Advisor and the Collector of Customs should not be discharged by one person."[3]

Sen. McCormick expanded on these recommendations to Secretary of State Charles Evans Hughes on Dec. 14, 1921. Stating that from his observations the Haitian people were wretchedly poor and that their merchants were suffering, he urged that the long pending loan be concluded without further delay. He suggested that American assistance was necessary in both the administration of Haitian justice and the support of education, while recognizing the silence of the treaty in these areas. He thought that tariff advantages similar to those granted Cuba were justified for Haiti in view of the responsibilities and authority assumed by the United States there. He concluded by asserting that, "Any policy on which we lay down in all probability cannot bear fruit in Haiti during the lifetime of one or two administrations in Washington. It is for this reason that the members of the Senate Committee hope that their report when it is published will serve to remove American policy in Haiti from the field of partisan controversy."[4]

The preliminary report was not well received by the Haiti-Santo Domingo Independence Society, especially in view of the fact that additional hearings in Washington had been promised so that its counsel might sum up the argument for the "prosecution." These hearings were held and a list of nine immediate and five ultimate demands were read into the record. If met, these would have resulted in the immediate termination of the American Occupation, punish-

3 Press release by the Select Committee on Haiti and Santo Domingo, undated. (Dec. 14, 1921?)

4 Sen. Medill McCormick to Secretary of State, Dec. 14, 1921.

ment of individuals found guilty of dereliction of duty and restoration of the democratic form of government presumed to have been functioning in Haiti prior to the landing of American Marines.

"By sheer force of arms, unjustified by any wrong committed or threatened to our own country, or to any other country, we have crushed the independence of another people, stamped out by force, and by clever devices bearing the pious sanctity of law, every vestige of native self government, every possibility of effective opposition to alien intrusion and domination. We have made of Haiti a vassal, a stepping stone under our southward marching absorption of Central America."[5]

The committee's final report, when it was published late in 1922, contained over 1,800 pages of transcript. Its conclusions were relatively brief, and its recommendations did not go much beyond those put forth in the preliminary report. The committee believed that American officials had exercised undue influence in the choice of a Haitian government and in the ratification of the 1915 treaty. It expressed its chagrin at the improper or criminal conduct of some few members of the Marine Corps in Haiti, but then went on to take a decided swipe at the Haiti-Santo Domingo Independence Society and *The Nation*. It found it to be its duty "to condemn the process by which biased or interested individuals and committees and propagandists have seized on isolated instances, or have adopted as true any rumor however vile or baseless in an effort to bring into general disrepute the whole American naval force in Haiti.

"The Committee wishes to express its admiration for the manner in which our men accomplished their difficult and delicate task."[6]

On July 7, 1922, Moorefield Storey, of Boston, an old friend of Secretary of State Charles Evans Hughes, wrote a personal letter to the secretary castigating him for his support of the administration's Haiti policy. In a postscript to the letter he added something that expressed the remarkable sentiment shared by those who remained unconvinced by the Senate Select Committee's report.

"If you quote to me the report of the Committee I can only say that it

5 Statement of Ernest Angell to Senate Select Committee on Haiti -Santo Domingo, June 15, 1922.
6 Report of the Senate Select Committee on Haiti-Santo Domingo.

was composed of Republicans and Democrats, that both parties are equally responsible for what has been done and is doing in Haiti, and that in these circumstances neither could criticize the other, and such a report as they made was inevitable, but from reading the evidence before the Committee it seems to me unwarranted."

Presumably, the only unbiased investigation possible would have to be that conducted by non-Republicans and non-Democrats (the editorial staff of *The Nation*?). Hughes replied sharply to this letter, thus alienating the first of many left-of-center friends and former associates who tried to use their personal acquaintance to influence him in political matters.

Sen. McCormick's wish for an end to partisan politics concerning Haiti was not to be. Others soon found that Haiti was an issue eminently political and proceeded to take advantage of that fact. Their role in keeping the controversy alive will be taken up later, after examining new developments in Haiti itself.

16

THE HIGH COMMISSIONER

... I am troubled to see that you have appointed Brigadier General Russell as a virtual dictator in Haiti. Was it not he who carried out the Wilson policy which you so uncompromisingly condemned? Is it not he whom the Haitians hold responsible not only as the agent of higher authorities in jamming alien rule down their throats, but for much oppression and cruelty in the process? I cannot but feel that this appointment will nullify all your efforts toward mutual understanding. I am therefore taking the liberty of enclosing the leading article in the Times Current History in the hope that it may present a view which will be useful in checking information you receive through official channels.

No one who knows you even as I as a mere boy knew you in Marion can doubt the essential humanity or your purpose and your loyalty to American traditions. May I urge that you now have an opportunity to re-establish the faith and confidence not only of Americans in our own traditions but of the weaker peoples in us? Few deeds would, I think, speak more clearly of peace and justice than restoration of full sovereignty as a precedent to whatever negotiations may seem desirable. Surely American help to the weak ought not to mean American coercion of them.

— Norman Thomas to President Warren G. Harding, March 7, 1922.

It had been the unconcealed hope of the Haiti-Santo Domingo Independence Society that one of its own would be named the High Commissioner to Haiti. An active campaign was waged to boost James Weldon Johnson for the assignment. The disappointment that was felt when Col. Russell was promoted to brigadier general and designated High Commissioner is therefore understandable. What *The Nation* did not know was that none other than Gen. Smedley Butler had been the Harding administration's first choice for

Col. John H. Russell, USMC.

the job. At the urging of some senators who had served on the Senate select committee, Col. Russell's name was substituted at the last moment. At no time does it appear that any consideration was given to filling the post with a civilian appointee because it was not thought proper to insert a civilian into the chain of command running from the Brigade Commander to the Secretary of the Navy.

Col. Russell, who was on leave in the United States when his appointment was announced, spent some weeks in the U.S. Department of State before returning to Haiti, where he participated in the drafting of a new organizational plan for treaty services and the Haitian government. The plan was to enhance American control while at the same time making it possible for the Occupation to be withdrawn without undue disruption in the functions of government.

Essentially, two parallel organizations were conceived, with all connections between the two going through, and only through, the High Commissioner. Haitian cabinet ministers would be limited to giving advice to their president, who would then, if he saw fit, transmit that advice to the High Commissioner. The position of cabinet officers would be one of great honor and dignity, but of little power and authority. The American officers actually conducting the affairs of the Republic of Haiti would be kept in the background and their Haitian counterparts pushed to the fore. At least at the outset, American influence in areas not touched upon in the treaty, specifically the courts and public education, would be exercised through the president.

President Harding announced Col. Russell's appointment on Feb.11, 1922. Exactly one month later, he arrived in Port-au-Prince on a battleship provided at the express request of the U.S. State Department. He immediately issued orders that implemented the plan worked out in Washington that gave the treaty organization something on the order of a military commander's staff. All correspondence addressed to the U.S. government, the Haitian government or to other treaty departments on matters other than routine henceforth would be transmitted through the office of the High Commissioner. All policy statements would come from him and him alone. All projects of law and budgetary proposals had to be cleared by him. He assumed responsibility for resolving matters in dispute between the Haitian government and a treaty official or between different treaty departments, with the U.S. State Department reserved as a court of last resort.

Inevitably, the new system tended to divorce advisers from their respective Haitian ministries, and made it impossible for an American official to line up on the wrong side of a dispute with the Haitian government. The treaty organization was thus immeasurably strengthened, but at the expense of a Haitian government that grew increasingly feeble.

Other changes growing out of the select committee's recommendations were made in Washington. The Navy's Bureau of Insular Affairs, which previously exercised a strong voice in Occupation matters, was abolished, and the Navy's role for all practicalities was reduced to the detailing of personnel and administration of the First Provisional Brigade of Marines. Authority was thus almost wholly centralized in the State Department's Division of Latin American Affairs. In January 1924, the two Occupation offices of financial adviser and general receiver were combined.

In the meantime, President Dartiguenave, taking advantage of the transitory provisions in Haiti's Constitution of 1918, failed to name the date for legislative elections in the even years of 1920 and 1922. He was encouraged in this by the U.S. government, which recognized that under existing conditions in Haiti not only was it improbable that any truly free election could be held (because of the ignorance and illiteracy of a majority of the electorate), but that legislative elections, if held without the exercise of undue influence by the Haitian gov-

President of Haiti, Louis Borno.

ernment and/or the Occupation, almost certainly would result in the election of a legislature largely anti-American in sentiment.

On Sept. 3, 1921, U.S. Secretary of State Hughes informed President Dartiguenave that the American government would not object should he decide to permit the election of his successor by Haiti's Council of State, as he was authorized to do in the absence of a legislature.

After certain parliamentary maneuvers, President Dartiguenave announced his candidacy for re-election, and though Gen. Russell reported there were strong feelings against him, it was generally assumed the Council of State, wholly appointed by Dartiguenave, would return him for another term. Thus it came as something of a surprise, especially to the opposition, when Louis Borno was elected president on April 10, 1922.

The successful candidate, who ran virtually unannounced, apparently bribed sufficient members of the council behind Dartiguenave's back.

Borno had a long if not distinguished career in the Haitian government. He served three times as minister for foreign affairs, and had signed both the Treaty of 1915 and the protocol, extending its life for an additional 10 years. He was considered strongly pro-American.

Immediately upon entering office, President Borno had the council enact long-deferred legislation necessary to put the Haitian Claims Commission into business. After competitive bids, a 6 percent loan of $16 million was awarded to National City Bank at a rate of 92.137. From the funds thus realized, more than $4 million went to the National Bank of Haiti (re-chartered in 1921 as a Haitian corporation wholly owned by National City Bank). More than $2 million was paid to the National Railroad to liquidate its large claims. Three French loans were refunded at a cost of more than $6 million, taking advantage of the depressed state of the French franc. Remaining funds were applied to public works and amortization of other loans.

An additional issue of $5 million in bonds was floated by National City Bank on the same terms as the $16 million loan. This was used to repay the unpaid balance on previous internal loans, and new awards made by the claims commission. More than 73,000 claims totaling just under $30 million were considered, and settlement was made at a rate of about 12 percent. Thus, Haiti's debts were at last put in reasonably manageable order.

The financial arrangements undertaken, however, were bitterly attacked in the opposition press. Wildly irresponsible charges were made against both the Borno government and American treaty officials. In August 1922, Gen. Russell issued a proclamation reminding the Haitian people of the nature of U.S. treaty obligations and re-stating the still existing responsibility of the military authorities under martial law to punish those found guilty of seditious or libelous acts. Particular attention was called to the penalties attached to violation of press laws.

Under his authority as brigade commander, on May 26, 1921, Russell had declared that, "While the freedom of the press and of speech are practically unrestricted, articles or speeches that are of an incendiary nature or reflect adversely upon the United States Forces in Haiti, or tend to stir up an agitation against the United States officials who are aiding and supporting the Constitutional Government of Haiti, or articles or speeches attacking the President of Haiti or the Haitian Government are prohibited and offenders against this order will be brought to trial before a Military Tribunal."[1] The language here was quite similar to that used by Adm. Caperton years before.

Prior to the Occupation, Hatian newspapers were held in check by fear of the existing government. Articles offensive to the government would result, under most regimes, in the people responsible being imprisoned or shot. After

1 Proclamation issued by the Brigade Commander, Port-au-Prince, May 26, 1921.

the American intervention, however, such action was impossible. The protection afforded to the newspapers resulted, however, not in a free press but in an unlicensed one characterized by the most irresponsible attacks on the Haitian goverment, treaty officials and American forces in Haiti. These often were of an offensive, personal nature. Redress through the Haitian courts ordinarily could not be expected, for in cases where American officials or the Haitian government were involved, the outcome of trial by jury was a foregone conclusion.

Russell's declaration was recognized in Washington as dangerous. When several Port-au-Prince editors were arrested and put on trial before the provost court, cries for help went out to the Haiti-Santo Domingo Independence Society. It found in the alleged suppression of a free press in Haiti the ammunition it had been looking for.

The greatest damage to the Occupation growing out of this was the alienation of Sen. McCormick, who was angered by the continued application of martial law in Haiti long after the need for it, in his opinion, should have disappeared. After the death of President Harding, the senator openly broke with the administration over its Haiti policy. On Jan. 13, 1924, McCormick wrote a long letter to newly inaugurated President Calvin Coolidge. He recited his grievences and recommended the sending to Haiti of a presidential commission to investigate and report what had occurred there in the two years since the Senate Select Committee had made its investigation. No satisfactory response having been made to his initiative, he introduced Senate Resolution 144 on Feb. 7, 1924:

"Whereas martial or military law was proclaimed in the territory of the Republic of Haiti by the Commander of the American military forces landed there in 1915; and whereas such military law continues effective throughout the territory of a friendly Republic by authority of the President of the United States; and whereas under such military law citizens of Haiti are liable to arrest by the armed forces of the United States and to trial before military tribunals of the United States nine years after military law was first proclaimed to the end that anarchy might be checked and civil law restored:

"Therefore, be it Resolved, that continuance of such military or martial law, and the liability of Haitian citizens, throughout the Republic, to trial before military tribunals of the United States, is undemocratic, unrepublican, and contrary to American ideals and the policies of Warren G. Harding, late President of the United States."

The phrase "undemocratic and unrepublican" caught on, and was cited countless times by delighted opponents of the Occupation. Martial law remained in effect in Haiti, but its application throughout the remainder of the Occupation was extremely rare.

Gen. Russell proved to be a skilled diplomat and an innovative administrator in the office of High Commissioner. In his supervision of the various treaty departments and in his negotiations with the Haitian government, the budget became his principal tool. Budgetary proposals and special appropriation projects were submitted concurrently to the High Commissioner and the Financial Adviser. Where differences of opinion existed between a department head and either or both of the senior reviewing officials, they were hammered out in executive session. Gen. Russell is never known to have authorized an expenditure not agreed to by the Financial Adviser, and his legal right to do so under the terms of the treaty was at least questionable. When agreement on money projects was reached, the respective treaty official took the matter up with his opposite in the Haitian government. Disagreements at this stage normally were resolved by conference between the High Commissioner and the Haitian President. When they failed to agree the issue was referred to Washington, but this rarely happened.

The core of the difficulty lay in the feeling on the part of Americans that money allocated for programs under Haitian administration by and large was money wasted, a feeling not entirely without justification. Consequently, since Americans controlled the collection and disbursement of public funds, the U.S. administered projects to expand at the expense of Haitian ones. Nowhere was this tendency more evident and nowhere did it stir up more acrimony than in the area of public instruction.

In April 1923, a protocol was signed which provided for the establishment of a technical bureau under American administration within the Haitian Department of Agriculture. This bureau was called *Service Technique de le l'Enseignement Professionel,* a title ordinarily shortened to *Service Technique.*

In February 1924, the Council of State passed a law assigning the Service Technique responsibility for improving Haiti's agriculture and for establishing a program of vocational education. In succeeding years increasingly larger appropriations were provided at the expense of sums voted for the Haitian-administered Ministry of Public Education. As touched on earlier, Haiti's educational system since the foundation of the Republic had been a disgrace. There were schools with no pupils, teachers who were illiterate and, on the part of the educationed elite, absolute disdain for the needs of the children of the peasant class. Consequently, there had never been any attempt to provide vocational schools even though countless critics had cited the crying need in Haiti for a skilled, working middle class. What public education under Haitian administration existed was for children of the elite, and schools nominally provided in rural areas were little more than sources of petty graft for their administrators. Schools for the city elite concentrated on the liberal arts and preparation for the professions, particularly law. Port-au-Prince during the Occupation may have had a greater proportion of lawyers than any city in the world. Manual labor was looked down on, and no self-respecting citizen would be caught dead carrying so much as a parcel in the streets. It is scarcely an exaggeration to say, as Occupation officials frequently did, that there were in Haiti few men who could saw a board straight or push a wheelbarrow correctly until shown how to do so by the Americans.

How deliberate the policy of keeping the Haitian peasant in ignorance is debateable. That it was deliberate, however, had been suggested for many years.

"Ah, if he could speak, our peasant! He would say very sensibily, 'I want light, I want progress, I want happiness. I am tired of my African superstitions from which you have done nothing to free me.... Do not pen me up in isolation of ignorance and distress which have made me a veritable serf. I am the true force, the real force which can save all. Have the loyalty, while it is yet still time, to permit light, progress and civilization to come to me.'" [2]

Another source of friction between American officials and the Haitian elite was the judicial system. Haitian courts at this time were of four classes.

2 Frederick Marcelin, "Une Evolution Necessaire," a pamphlet published in Paris, 1898.

At the lowest level were the *Tribunaux de Paix*, or police courts. There were about 125 of these and they were notoriously corrupt, venal and inefficient. A favorite story of the Occupation concerned the case of a man who had been arrested for stealing the door off a house. Even though he was apprehended with the door still in his possession, he was acquitted by the *Juge de Paix,* who by chance happened to be the man's cousin.

Above the police courts were 11 Courts of the First Instance, and over these was the *Cour de Cassation.* The common complaint of American treaty officials was that it was absolutely impossible for an American or an American-administered department of the Haitian government to get impartial treatment in a Haitian court. The judicial system was blamed for discouraging the investment of foreign capital in Haiti. The courts were condemned for awarding excessive and unwarranted judgments in suits against the government, judgments which the Financial Adviser frequently refused to pay. This, while it doubtless saved the Haitian government money, did nothing to improve respect for law.

President Borno followed the lead of his predecessor and failed to name the date for legislative elections in 1924 and 1926. In the latter year, again with the blessing of the State Department, he was elected to a second four-year term. As opposed to his first election, when a scuffle broke out and the voting urn was turned over and destroyed, the balloting in 1926 was uneventful. His re-election by the Council of State, however, was condemned by the political opposition.

In 1927, Gen. Russell determined that the ultimate success of the Occupation required revisions of the Haitian Constitution. It was considered too risky to have the revisions approved by a specially elected National Assembly, as provided for in the Constitution. It was thought this would result in bringing together active obstructionists. Therefore, it was decided to employ the procedure followed in 1918. The Council of State would approve the revisions and then they would be submitted to the people in a referendum.

The changes desired by the High Commissioner concerned the Haitian judiciary. It was proposed to suspend, for a period of one year, the irremovable nature of judicial appointments. This, combined with the president's power to appoint judges, would permit replacement of those thought antagonistic to the government. The terms of judges were reduced from life to seven years, except for the Cour de Cassation, where a 10-year term was specified with life tenure after a justice had served there for 25 years. The Cour de Cassation would also function as a court of appeals, thus providing the government with an avenue of redress from unfavorable rulings in lower courts.

President Borno and the Council of State had suggestions of their own. The Gendarmerie would be renamed the Garde, a more descriptive title in view of its expanded function. Trial by jury no longer would be guaranteed to violators of press laws. The term of office for senators would be changed from four to six years, and the president from four to seven years. The legislature (including the Council of State acting in the absence of a legislature) would have the exclusive power to interpret the Constitution.

The last two recommended changes were strongly opposed by the U.S. State Department. After considerable back and forth it was agreed that interpretation of the Constitution and the law would be a function of the Cour de Cassation, following the American model. The president's term of office would be fixed at six years, with no eligibility for re-election. It was understood, however, that incumbent President Borno would be eligible for re-election to a final term of six years in 1930. This was important, for the man thus elected would hold office in the crucial final years of the Occupation which, under the terms of the treaty, would expire in 1936. This point was not missed by the opposition.

On Jan. 10, 1828, the proposed amendments to the Constitution were adopted in public referendum by a vote of 177,436 to 3,799. This was understandably jeered by those who wished to know why, if the people were capable of voting for amendments they could not read, they were not thought capable of electing representatives to the national legislature.

As the decade of the 1920s drew to a close, the military forces supporting the Occupation had sunk to a skeletal brigade of Marines numbering scarcely 500 men. Another 100 Marines served in the Garde d' Haiti. The treaty organization under the High Commissioner consisted of five departments: The Garde, the Financial Adviser-General Receiver, Public Works, Public Health Service and the Technical Service of Agriculture. Americans employed in the various departments numbered about 250.

The Haitian public debt, the structure of which had been a prime factor in the 1915 intervention, had been reduced by half and the balance re-financed on favorable terms. Foreign claims had been settled. Steps had been taken to disentangle Haiti from its disastrous railroad concessions. In all of these complicated financial arrangements, no breath of scandal had touched American officials, to the deep chagrin of those who searched carefully but in vain for a Haitian "Teapot Dome" or a new "Consolidation Affair." Economic progress was slow, painfully slow, for with the exception of salaries paid to Navy and Marine personnel serving in Haiti, the United States did not pay any of the costs of the Occupation. No financial aid was given.

Visible improvements were made in sanitation, roads, hospitals, prisons, water systems and schools. The annual reports filed by the High Commissioner were filled with statistics and photographs documenting material progress. To those familiar with condidtions in Haiti both before and after the intervention, the transformation was said to be remarkable.

The prolonged failure to introduce representative government was painfully evident to all, however. It was clear the Haitian people in 1929, almost a whole generation after the intervention, were no more prepared to govern themselves than they were on that day in 1915 when Adm. Caperton arrived to see body parts of their late president paraded through the streets of Port-au-Prince.

17

"THERE IS NO POLITICS IN THIS"

The United States invaded the Republics of Haiti and Santo Domingo and is holding them in subjugation through martial law for no other reason than to allow American corporations to grab the rich resources of the two republics and to exploit them, according to Sen. Borah of Idaho. The Senator spoke last night at a mass meeting in Carnegie Hall called by the Hayti-Santo Domingo Independence Society to protest against the continued occupation of the island republics by the United States....

The auditorium was crowded from stage to topmost balcony and Sen. Borah was applauded and cheered.... "There is no politics in this," announced Sen. Borah. "The Democrats went in and the Republicans are staying in.... If I had thought that this occupation was only a temporary affair I would not be here to talk to you tonight. I am firmly convinced that we are there to stay for all time unless public opinion rises to a pitch to change conditions.... Not only are we concerned with the welfare of the natives of these republics but we are concerned with the honor and integrity of our country and our people...."

The New York World, May 2, 1922.

In April 1922, the Foreign Policy Association of New York published a legal brief titled "The Seizure of Haiti by the United States," which was endorsed and distributed by the National Popular Government League of Washington, D.C. It was reportedly authored by 24 distinguished lawyers, including future U.S. Supreme Court Associate Justice Felix Frankfurter. The brief argued that continued presence of U.S. military forces in Haiti was a violation of well-recognized American principles, and that this was not only contrary to international law but also to the U.S. Constitution. The work concluded with

Secretary of State Charles Evan Hughes, left, and Sen. William E. Borah, right.

a call for the immediate abrogation of the Treaty of 1915, the holding of free Haitian elections and negotiation of a new treaty "upon such terms as shall be mutually satisfactory to both countries and by the methods that obtain between free and independent states."[1] The legal brief was a strong one, and the qualifications of those endorsing it impeccable.

The intervention was defended by Dr. Carl Kelsey, a respected scholar, in a paper presented to the American Academy of Political Science in 1922 titled "The American Intervention in Haiti and the Dominican Republic." In effect, Kelsey told critics of the Occupation to "check their premises."

"There never has been any election in Haiti. There never has been any democracy in Haiti. It is a perfectly foolish use of language to talk as if there had been. Haiti traded a slave system under white slaveholders for a slave

1 Foreign Policy Association, "The Seizure of Haiti by the United States," Pamphlet 8, New York, 1922.

system under mulatto owners, and they have run under a slave regime from the first up to the present time....

"A century ago men sensing an idea a little beyond their powers of expression spoke glowingly of 'individual rights' as if they arose and existed apart from society. We know today that rights flow from society and are determined by it. To society the individual is responsible and when the commands of society are violated, the individual is punished; that is, his rights are limited. Society judges the individual by his actions and not by his size. It recognizes that individuals differ and that the rights granted must be proportionate to the sense of responsibility developed by the individual. The insane man must have a guardian. When we deal with defective persons we do so not to punish them but to assist them and to protect others, that is, society....

"The individual is most affected by the actions of his neighbors. If a man on the adjoining place begins to shoot indiscriminately with a high-powered rifle, I am immediately involved regardless of whether he intends to do me harm or not. In an organized society I invoke the law. Under frontier conditions I handle the problem myself. In the present state of the world organizations, we must follow the program of the frontier."[2]

The philosophical battle was joined. The "24 lawyers" relied on the letter of the law and testimony in the published transcript of a Senate select committee to show that the United States had broken the law. Dr. Kelsey provided much-needed support for the U.S. State Department in its attempt to justify admittedly illegal actions in Haiti on the grounds that it was dealing with neither the constitutional government of Haiti nor representatives of the Haitian people, but rather with irresponsible political leaders who had come into prominence as the leaders of armed uprisings. In all such battles, however, the advantage lies with the side that has the initiative, and the initiative never left the opponents of the Occupation.

On April 29, 1922, Gen. Russell informed the U.S. Secretary of State that Dr. Gruening, on behalf of the Haiti-Santo Domingo Independence Society,

2 Carl Kelsey, "The American Intervention in Haiti and The Dominican Republic." The American Academy of Political Science, Philadelphia, 1922.

had asked the Union Patriotique to raise the sum of $12,000 in gold as its share of the expenses associated with the holding of a mass meeting in New York. The meeting was jointly sponsored by the U.S. Foreign Policy Association to protest the election in Haiti by its Council of State. Dr. Gruening promised that Sen. William E. Borah would be one of the speakers.

Subscriptions were solicited in Port-au-Prince, and Gen. Russell was told that outgoing President Dartiguenave gave $50 and his cabinet $250. Press accounts of Sen. Borah's speech in Carnegie Hall circulated widely in Haiti.

Members of the National Civic Federation were in the audience for the May Day speech. Afterwards, Louis Marshall declared that Germany did nothing in Belgium to equal the atrocities committed by Americans in Haiti and Santo Domingo. Ralph Easley, chairman of the federation's executive council, which included such well-known figures as Marshall Field and Samuel Gompers, wrote immediately to Secretary of State Charles Evans Hughes for material on which to base a reply. "It goes without saying," Marshall wrote, "that the policy originated by the late administration, continued and defended by the present one and sustained in a report made by a commission of Democratic and Republican senators, cannot be open to any such vicious charges as are made by this Foreign Policy Association and its radical adjunct, the National Popular Government League. In fact, the public would generally assume that such a policy must be fairly just."[3]

In the same mailing with this letter to Hughes, another of a completely different character was sent by Marshall, an old friend of the secretary. Marshall bitterly complained about the treatment afforded the "24 lawyers" at a meeting at the U.S. State Department.

"... I was therefore not only astonished but stunned when you began your remarks with the announcement that you regarded the statement that had been made [the seizure of Haiti by the United States] as most inadequate and one-sided, and closed with the paragraph: 'You need not for a moment suppose that any of these matters are not engaging the attention of the Department; we have very full information, in light of which I am bound to say the statements

3 Robert Easley to Charles Evans Hughes, May 2, 1922..

you have made here are extremely inadequate.'

"I had always fancied that the right of petition existed; that it was the privilege of a citizen to present facts, as he understood them, in an orderly and respectful manner; that he was not to be subjected to public contumely because of that fact, nor that an oral pronouncement such as yours would be followed by the publication of the full text in the newspapers....

"It is quite certain that those who believe in the doctrine of manifest destiny will contend that our Government has a moral right to dominate over inferior nations, to those who desire an opportunity to exploit the rich lands of Haiti and to attend to her finances, will agree that we have done no wrong. That may be the other side of the question. But I mistake the American people if they will ever accept such doctrines...."[4]

Marshall may or may not have been prepared for the reply fired back by Secretary Hughes. The rebuttal of his grievances was devastating in tone:

"... I said that the statements which you and others made to me with respect to the relations of the United States and Haiti were inadequate and one-sided. I described these statements in this way, first, because I felt this was a just description, and, second, because the circumstances made it necessary that they should be thus described.

"Of course you have the right of petition. I recognized it in receiving you, and certainly you fully exercised it. But this does not mean that a public officer is to be deprived of the right to put in a caveat against the assumption on the right of the public that statements submitted should be accepted. If you bombard a public officer in this way, he is at least entitled to avoid the misinterpretation that might easily be put upon his silence.

"I confess that I am amused at the humor, apparently unconscious, of your complaint that my remarks were made public. Your brief, which I had already read, and the striking extracts from the oral statements made, had already been given to the press, as I knew at the time, subject to release after the interview. In short, your staged a public demonstration in order to ensure publicity for your assertions and demands, and surely you have no right to object to my pub-

4 Louis Marshall to Charles Evans Hughes, May 2, 1922.

lishing a rejoinder that I thought your statements inadequate and one-sided....

"I recognize your rights, although I do not commend your attitude. If you were really desirous of aiding the Administration in its handling of a difficult and highly important situation, you would have come to me — and our old friendship would seem to have suggested that course — in order to secure such information as could properly be given to you. You might have learned that the Administration had been devoting itself with the utmost earnestness and solicitude to this problem. Instead you saw fit, exercising your undoubted right, to present your peremptory demands.

"I shall continue to deal with this matter and to advise the President according to my best judgment, as fully conscious of the demands of justice and of the Nation's honor as you yourself can possibly be. We shall, of course, seek in every possible way to conserve the welfare and independence of the people of Haiti and to end the military occupation as soon as can properly be done. We shall make such public statements from time to time as, in this position of graver responsibility, may be deemed compatible with the public interest."[5]

This letter quieted Louis Marshall and the authors of the legal brief for a moment, but the pot kept boiling from fires set elsewhere. In 1922, Sen. William H. King, recently a member of the Senate Select Committee, introduced an amendment to the military appropriations bill declaring that no funds should be used for the maintenance of Marines in Haiti or the Dominican Republic. The amendment was defeated, but Sens. King and Borah took the occasion to make ringing denunciations of U.S. policy in the two countries. On April 26, the Joint Committee of the Federal Council of the Churches of Christ in America delivered a memorandum to President Harding protesting his policy in Haiti. This was but one of many church groups that took similar stands.

Protests in the United States encouraged more open actions in Haiti, and the Gendarmerie had its hands full keeping the situation under control. "Macandalism" entered the vocabulary of the opposition in Port-au-Prince. It referred to a black man named Macandal, who in the time of the revolution against the French, being skilled in the use of herbs and poisons, succeeded in killing

5 Charles Evans Hughes to Louis Marshall, May 5, 1922.

many whites by subverting their servants. Thinly veiled threats appeared in the opposition press. Some, such as the following, were open invitations to murder:

"... Louis the Pious [President Borno] is crazy. Even his most intimate friends no longer can hide the fact. We must add that he is not just crazed — he is furiously and dangerously insane.... Has not the moment arrived to feed a sausage to Louis?"[6]

Dozens of examples could be drawn from opposition papers in Port-au-Prince in this period encouraging violence against Americans and government officials. The Occupation was virtually helpless to suppress such activity, not only because of the disinclination of Haitian courts to convict in such cases but also because of anticipated reaction in the United States to any infringement on Haiti's "free" press.

In October 1924 a campaign was launched in the French press criticizing the Occupation. The occasion was an address by a member of the Union Patriotique, Dante Bellegarde, in Lyons before the International Union of Associations in favor of the League of Nations. The address used the principle of self-determination for the people of small nations as a point of departure. When Bellegarde returned from France he carried a mandate from the *Ligue des Droits de l'Homme et du Citoyen* to form a local branch of that association in Port-au-Prince.

In March 1925 the French prime minister was petitioned to bring up the question of the U.S. military occupation of Haiti before the next session of the Council of the League of Nations.

The Universal Negro Improvement Association, according to Port-au-Prince newspapers, asked in a resolution unanimously voted by "the negroes of five continents" that the last Sunday in October be celebrated as "Haiti Independence Day." Widespread support for Haitian independence flared in the Latin American press. Rumors of imminent American withdrawal circulated.

At Gonaives, a general of the old Haitian army rode into town to take charge as the *Commandant de Arrondissement.* He had heard out in the hills that the Americans were leaving. He wore his old uniform.

6 *Courrier Haitien*, Port au Prince, June 12, 1924.

In the United States, the Universal Negro Improvement Association's Marcus Garvey, whose "Negro World" was a scheme for a new colored nation in Africa, joined forces with James Weldon Johnson, Dr. Gruening, Oscar Garrison Villard and other charter members of the Haiti-Santo Domingo Independence Society in a plan to use disruptions in Haiti as an embarrassment to the Coolidge administration during the U.S. national elections in 1924. Instructions were sent to the Union Patriotique to start an active propaganda campaign throughout Haiti, and to print in the opposition press abusive and defamatory articles against Coolidge and the Occupation, with a view to securing arrests for violation of press laws. This would then be used as political fodder in U.S. elections.[7] These instructions fell into the hands of the U.S. High Commissioner to Haiti. The abuses were numerous but no arrests were made.

In May 1925, Sen. King called at the U.S. State Department with Pierre Hudicourt, a prominent member of the Union Patriotique, and expressed himself at great length concerning Haiti. Sen. King stated that, if he were President of the United States, the Marines would be withdrawn from the island the next day. He said that if the Haitians desired bloodshed and revolution, it was their privilege. He said American policy in Haiti would bring about a revolution there, and he would support it fully.[8] Hudincourt listened to this with undisguised satisfaction. Sen. King thus stole a march on Sen. Borah and became the American patron saint of the Union Patriotique. A photograph of Sen. King seated amid a half-dozen leaders of the Haitian opposition was sold on the streets of Port-au-Prince to raise funds to fight the Occupation.

In the spring of 1927, Sen. King announced plans to visit Haiti and this caused President Borno to order the Gendarmerie to prevent his entrance into the republic. The opposition press headlined a letter from the senator to the Union Patriotique on March 31 extremely critical of the Haitian leader, the U.S. State Department and the High Commissioner.

In the summer of 1925 there had been considerable agitation in Haiti and the United States concerning Haitian elections, 1926 being an "even" year

7 Russell to Sec. of State, No. 435, Oct. 30, 1924.

8 Internal memmorandum, Div. of Latin American Affairs, May 22, 1925.

and one in which Borno's first term would end. Not even Mrs. Coolidge was spared a large influx of mail on this subject. In Port-au-Prince a member of the Union Patriotique announced his candidacy for president and ran on a platform calling for 15,000 primary schools; 1,000 forts in the mountains, trenches provided with high-caliber and long-range cannons; an aerial and naval fleet; purchase of the Banque National de la Republic d'Haiti free from the control of the National City Bank of New York; negotiation of a loan of up to five billion dollars worth of gold on the Belgium market; diminution of expenses and an increase in salaries."[9]

Emily Greene Balch.

The U.S. High Commissioner commented that it would be difficult to find a platform that would more appeal to ignorance and prejudice. As already noted, however, President Borno did not set a date for legislative elections in 1926, and was re-elected to a second term by Haiti's Council of State.

In the wake of the announcement that there would be no legislative elections that year, President Borno's popularity fell to a new low. On Jan. 3, 1926, he presided at the disinterment and reburial together in the same grave of the remains of Dessalines and Pétion. There was no applause following his speech. The claque's shouts of *"Vive Borno!"* were met by laughter.

On Feb. 19, 1926, a committee sponsored by the Women's International League for Peace and Freedom (WILPF) sailed from the United States to

9 Russell to Sec. of State, No. 614, Aug. 27, 1925.

Haiti to "look into the conditions there." The committee was nominally headed by Emily Greene Balch, a well-known sociologist, and it consisted of two representatives of the American branch of the WILPF, two representative American black women, a representative of the Fellowship of Reconciliation and an Amherst College economics professor representing the Foreign Service Committee of the Society of Friends. The professor was Paul H. Douglas, the only man in the group, who later became a decorated combat Marine in World War II and a distinguished U.S. senator.

On arrival on Feb. 24 in Port-au-Prince, the committee members were warmly welcomed in the pages of the *Union Patriotique.* They called on the High Commissioner and assured him their mission was purely philanthropic with no political agenda. Gen. Russell, however, sized them up as "internationalists and pacifists." He reported that they showed little interest in the Occupation's achievements and eager to uncover past mistakes. Gen. Russell assured the committee that the Occupation's work was an open book and he had nothing to conceal. On the contrary, the general said, he was exceedingly proud of what had been accomplished in assisting the backward nation, and that if the committee came with open minds they too would be proud as Americans. But if they came with fixed principles and preconceived ideas, they might as well return home at once.[10]

Though friendly assurances were given during the visit with Gen. Russell, Miss Balch soon thereafter angered him. She literally flew into the arms of Jane Adams, who made an overnight stop in Port-au-Prince on the cruise ship *Mount Royal* with a wholly unsubstantiated story that the High Commissioner had prevented her racially integrated party from registering at Haiti's Hotel Montaigne. Adams, of Hull House fame, was also the president of the Women's International League for Peace and Freedom.

Gen. Russell summoned Miss Balch and Paul Douglas to the American Legation and extracted an apology for their "error," but there was no retraction.

While in Port-au-Prince the committee called on foreign representatives in Haiti to get their views on the Occupation. At no time did they take pains to

10 Russell to Sec. of State, No. 735, March 13, 1926.

conceal their opposition. Prior to returning home, they called again on the High Commissioner and urged him to use his influence to end harassment of the press and to secure free and open elections.

Once back in the United States, the committee members collaborated on a report, edited by Miss Balch, titled "Occupied Haiti: Being the report of a committee of six disinterested Americans representing organizations exclusively American who, having personally studied conditions in Haiti in 1926, favor restoration of the negro republic." The report urged that an official U.S. commission be sent to Haiti to begin the work necessary for a transition of the government to Haitian control, and the eventual end of the Occupation. On the whole, "Occupied Haiti" presented the most balanced and rational attack on the American Occupation of any produced in this period. It found that from the point of view of selfish American interests, there was little to justify continuing the Occupation. American investments in Haiti were not profitable. The American military presence was damaging U.S. relations with Latin America. Military control of civilian affairs was the worst possible model for Haiti. The standard criticisms of the Occupation were leveled: Marine rowdiness, racism, contempt for free elections and interference with freedom of the press.

The committee's major recommendations were: Withdrawal by 1936, appoint a commission to investigate and report on the situation, legislative elections and demilitarization of American administration of the country while it lasted, "Haitianization" and gradual withdrawal of the U.S. Marines. The U.S. government eventually adopted all of this. When the report drifted into the puritanical, however, it seemed a bit ridiculous. "Disquieting rumors are abroad of pressure in favor of establishing gambling palaces as an attraction to tourists. We emphatically protest against such a plan."[11]

Working drafts of the report and a separate article prepared by Paul Douglas for publication in *Foreign Notes* were sent to the U.S. State Department for comment and possible correction. None were provided.

11 Emily Greene Balch, ed. "Occupied Haiti," Women's International League for Peace and Freedom, New York, 1927.

On May 13, 1927, the enlisted strength of the Marine brigade in Haiti was reduced to 452 when a battalion was withdrawn for service in Nicaragua. Gen. Russell requested their return or replacement as soon as possible in view of likely disturbances when constitutional changes were announced in the fall. Authorized brigade strength was left at about 500.

No elections were held in 1928. The high commissioner reported that President Borno finally and definitely agreed to allow legislative elections in January 1930. By that time it was thought that all important laws would have been drafted and voted on by Haiti's Council of State, and all governmental machinery, with the exception of the legislature, would be in place. Upon receipt of this optimistic report, U.S. Secretary of State Frank B. Kellogg wired Borno his personal congratulations.

A crucial year laid ahead, one that would prove decisive for the Occupation.

18

THE DISORDERS OF 1929

If you hear the politicians groan it is because they are expiating their lack of humanity and presumption, and their bad faith. They never loved you; they never wished the best for you. They, who forged all your miseries, all your troubles, now mention you out of the purest hypocrisy in their desperate appeals to the misguided philanthropy of the Kings and Borahs. All the philanthropists of the world could not save a nation in disorder, whose children close their ears to every word of reason....

In this country it is almost impossible to do good. Our most distinguished men, for the most part, have apparently sworn to spend their lives lying, deceiving and leading into error the thousands of illiterates purposely kept in that degrading condition for a century....

Let the whites withdraw tomorrow, and the day after tomorrow the country will be put to fire and sword. In spite of the lesson of fully 14 years, are we able to put our hands on fifty Haitian Statesmen?

— *L'Essor,* editorial, Port-au-Prince, April 5, 1929.

President Borno's "decision" to hold legislative elections in 1930 was not publicized and was known only to himself, the High Commissioner, and the U.S. State Department. In the spring of 1929 he began to have second thoughts on the matter, saying he had not adequately realized the "tremendous upheaval" that undoubtedly would be caused by legislative and presidential elections in the same year. Gen. Russell and certain State Department officials shared these sentiments. The Haitian administration elected in 1930 would be in office when negotiations for a possible renewal of the Treaty of 1915, due to expire in 1936, were held.

In March 1929, the head of the Latin American Affairs Division, Dana G.

Munro, informed Under Secretary of State Francis White that any election held would be a farce under existing circumstances in Haiti, and that "any Congress which might be elected, even though the Borno administration dominated its selection, would inevitably be a serious obstacle to further progress."[1] White agreed, and so advised Secretary of State Henry L. Stimson, who said calling legislative elections was a Haitian responsibility.

The opposition thus was left in the dark about President Borno's intentions (he had until Oct. 10 to announce the date of the election if he decided to hold one), but they suspected the worst. Agitation continued in the press.

On April 22, 1929, three opposition journalists were tried and convicted for violation of the press laws, specifically for denouncing President Borno as a traitor and the Archbishop of Haiti for racial prejudice and pro-American sentiment. In the poisonous atmosphere of Port-au-Prince, extraordinary precautions were taken at the trial to prevent an outbreak of violence. One American officer, one Haitian officer and a detective were placed in charge of the prisoners, with additional instructions to remain in the immediate vicinity of the judge for his protection. Ten members of the newly organized Garde d' Haiti, unarmed except for police clubs, and 10 plainclothesmen were stationed in the courtroom. Three Americans - two non-commissioned officers and 16 privates, all members of the Garde - were posted in the street in front of the courthouse. One American officer was at the front entrance, one at the rear entrance and one patrolled on a motorcycle. The commissioned and noncommissioned officers carried revolvers. The other members of the Garde were armed with police clubs.

Under the courthouse and not in view, one noncommissioned officer and 10 privates armed only with police clubs were held in ready for any emergency. The fire department had a steamer in reserve at the Garde garage, 200 yards distant. Another American officer was in charge there. This precaution was taken should it be necessary to disperse a crowd by hitting them with high-pressure streams of water. Four American officers were stationed in the hallway outside the courtroom.

1 Internal Department Memorandum, March 16, 1929.

At the police station, 400 yards away, five noncommissioned officers and 15 privates, armed respectively with revolvers and rifles, were held in reserve under the command of an American officer. The 12th Mobile Company of the Garde was held in reserve at the Caserne.

The Garde's commandant said that, of this entire force, only 12 uniformed members were inside the courtroom. Twenty-one were in view outside of the courthouse and all others were held in reserve and out of sight. This was done, he said, so that no charge of intimidation could be entertained.[2]

The journalists on trial were Jacques Roumain, George J. Petit and Elie Guerin. The defense attorney gave a ringing speech in which, to the cheers of the spectators, he called President Borno an "ass with three heads." At the conclusion of his speech, the prisoners embraced their lawyer by leaning over the prisoner's box and one, Roumain, made as if to step over it. He was restrained by an American officer and in the struggle the box broke under their weight and they both fell to the ground.

The courtroom was in extreme disorder. Part of the audience surged toward the bench. People were shouting emotionally and the *Commissaire du Gouvernement* fled from the room. The judge sought refuge under his desk and later disappeared. He returned after order was restored. Three women in the middle of the room shrieked.

The American officer who had fallen managed to get back up, only to be rushed by Roumain, who attempted to throttle him. The Haitian chief of detectives finally subdued the irate journalist by striking him twice on the head with a nightstick, leaving bloody scalp wounds. American officers carried Roumain to the judge's desk and wrapped his head with their handkerchiefs. An angry crowd closed in on the chief of detectives. He quieted them by drawing his revolver. All three prisoners were led out of the courtroom to a police van.

The version of this affair that circulated in the opposition press claimed that Roumain had no intention of starting a demonstration, and that his sole idea in trying to leave the box was to kiss his sister farewell.

On Oct. 5, 1929, President Borno announced that there would be no elec-

2 Memorandum for the High Commissioner from Commandant, Garde d'Haiti, May 1, 1929.

tions for the legislature in 1930, and consequently the Haitian Council of State would elect the next president. A storm of protest rose in Haiti and the United States. A National League for Constitutional Action was formed in Port-au-Prince, and in Washington, The People's Lobby, whose slogan was, "To Fight for the People We Get and Give the Facts," demanded that the U.S. Secretary of State take action. He steadfastly maintained, however, that it was a Haitian decision whether elections were to be held or not.

It is possible this storm might also have blown over had it not been for a "strike" of students at the Service Technique's Central School of Agriculture at Damien, on Oct. 31, 1929. It had been traditional in Haiti to provide promising students with a *bourse* (scholarship) to encourage them to continue their studies. Most of these were given to children of the urban elite, who had enjoyed a vastly superior primary education in addition to other advantages over the poor. In 1929, however, it was decided to reduce the $10,000 appropriation for the bourse by 20 percent, and to use the $2,000 saved to pay wages to rural students working on Service Technique demonstration farms and experimental stations. Country students welcomed the program. But it enraged city students who, trained from childhood to regard manual labor as degrading, saw their class privilege eroded. The strike at Damien spread to other schools, including those at the primary level. At Cap Haitian it was reported that children no longer saluted the national flag at morning exercises, and they were disrespectful of their teachers. At Jacmel, during a student parade, the only banner carried was a red flag with a green serpent. When the school band struck up the national anthem, student leaders put a stop to it.

Opposition leaders fanned the unrest and slowly converted it into a general protest against the Borno administration and the American Occupation. On Dec. 4, 1929, insolent customs house employees in Port-au-Prince attacked an American supervisor and threw typewriters and inkwells on the floor. A mob forced its way into the office of the Financial Adviser and demanded higher wages. In these circumstances, Gen. Russell felt compelled to issue the following proclamation:

"The United States Forces in Haiti are engaged in aiding and supporting the constitutional government of Haiti, and are your friends. By their efforts and those of the Garde d'Haiti, peace and tranquility have been established throughout your land for many years, permitting you to conduct your business and earn an honest living.

"Certain agitators are now endeavoring to foment trouble. It therefore becomes necessary to place in towns the power and authority of martial law, which has during the past few years been inoperative.

"You are also informed that articles and speeches of an incendiary nature or those that reflect adversely upon the United States forces in Haiti or tend to stir up agitation against the United States officials who are aiding and supporting the constitutional government of Haiti, are prohibited, and offenders against this order will be brought to trial before a military tribunal.

"From the promulgation of this proclamation, all inhabitants of the cities of Port-au-Prince and Cape Haitian will remain in their houses from 9 p.m. until daylight."[3]

Gen. Russell followed up his order with an urgent request for an increase of 500 men in the strength of the Marine Brigade until after the inauguration of a new president in 1930. This was received with the greatest dismay in Washington. Secretary of State Stimson sent Gen. Russell a "Rush, Double Priority" wire requesting that he hold his order pending receipt of his full views on the matter. But it was too late, even if circumstances had permitted less drastic action.

Stimson's long personal message to Gen. Russell arrived a few hours after the proclamation of renewed martial law and curfew. Stimson said the Navy had been instructed to place the USS *Galveston* and its Marine detachment, then at Guantanamo, at Gen. Russell's disposal, but that he hoped it would not be necessary to land additional forces.

"Having said that, you will be supported in your final responsibility of protecting life, I nevertheless think it proper to say from this distance it does not seem the situation requires or will be best served by a display of Marine forces in the outlying portions of the island. It seems to me that the nub of your situation rests in the question of the loyalty of the native constabulary,

3 Proclamation of the First Provisional Brigade of Marines in Haiti, Dec. 4, 1929.

and I hope you will be careful not to discourage that loyalty by an appearance of distrust or of supplanting it with white forces. Under white officers such as you have, the general loyalty of black troops is usually proven against even local dissatisfaction or mutiny. In whatever manner you ultimately decide upon I trust you will give the loyalty of your constabulary a thorough and fair test. Even if you deem it necessary to use more actively than hitherto the present force of Marines, I seriously question the wisdom of the proclamation as to martial law. To me it seems that the benefit of such a proclamation in an illiterate population like Haiti is outweighed by the unfortunate effect produced in the United States, particularly as you state that martial law was already in force. I therefore sincerely hope that you may soon be able to greatly moderate or withdraw that proclamation. In so doing I suggest that you make it clear that the present issue is not between the people of Haiti and the forces of the United States but between the agitators against the lawfully constituted authority of Haiti supported by the United States.

"In summary I sincerely hope that you may find it possible to protect life by withdrawal of Americans in exposed places rather than by extending the use of the present Marine forces or calling for additional Marines. The responsibility is, however, upon you and will be respected."[4]

Gen. Russell's attempt to meet Secretary Stimson's desires in the crisis was overtaken by events, for the next day a skirmish between a Haitian mob and a Marine patrol at Aux Cayes left six dead and 28 wounded, all Haitian. Four of the wounded later died. In these circumstances the 500 additional Marines that Gen. Russell had requested were immediately sent from the United States.

The situation at Aux Cayes began building in late November when opposition emissaries from Port-au-Prince arrived to organize a school children's strike. Stevedores were persuaded to join in the disturbances, and a small detachment of Marines arrived to quiet things down. Agitation increased, however, and spread to peasants in the surrounding countryside who already were angered by a drastic fall in the price they received for their coffee crop (caused by world market conditions). A mob of about 1,500 gathered at the outskirts of the town

4 Secretary of State to Russell, despatch, Dec. 5, 1929.

and advanced on the center. They were halted by 20 Marines with rifles and a machine gun. The mob leaders were allowed to enter the town. There they met with opposition politicians. They returned and demanded release of three prisoners from the town jail. This was refused and rocks pelted the Marines.

In what may have been the first American use of air power for psychological purposes, an old Marine scout plane flew over the mob and dropped several small bombs in the harbor. But this only briefly halted the mob's advance. Shots were fired over the heads of those in the front ranks. When this also failed to stop the mob's approach, more shots were fired resulting in 34 casualties, and instant dispersal of the mob.

The affair at Aux Cayes made headlines in the American press and evoked yet another round of demands from the anti-Occupation lobby for action by President Hoover to investigate and end the "mess" in Haiti. At an "off the record" press conference, Secretary Stimson referred to the Haitian situation as "another baby he found on his doorstep."[5]

5 Memorandum of a Press Conference by Secretary of State, Dec. 6, 1929.

19

THE PRESIDENT'S COMMISSION

Sur la vaste mer un bateau est ballotté par la tempete. Le patron envoie le mousse en haut du mat pour plier un voile. Le mousse crie "Je vais tomber, le tete me tourne." Le patron lui dit: Ne regarde pas en bas, regarde les étoiles.

— Francois Dalencour, *Le Sauvetage National Par le Retour à la Terre.*

A veritable flood of protest and advice concerning the situation in Haiti converged on Washington even before the worst of the news from there became known. On Oct. 3, 1929, the Union Patriotique addressed an open letter to President Hoover criticizing past American influence concerning President Borno's election and re-election (said to be unconstitutional because he was not born of a Haitian father); the electoral laws enacted; reform of the judiciary; taxes; immigration; and the budget.

The letter ended with a fervent demand for immediate legislative elections. A telegram from the Save Haiti League of America quickly echoed the Union Patriotique's letter and urged that President Hoover force Haiti to have elections. The Labor Open Forum of Rochester protested the continued presence of American Marines in Haiti.

The American Friends Service Committee of Philadelphia urged the appointment of a commission to investigate affairs in Haiti, and recommended that Paul H. Douglas serve as a member. Douglas advised President Hoover that no appointment to the proposed commission should include residents of Haiti because they would be unable to give unbiased statements for fear of reprisal from President Borno. He also warned that American officials in the Occupation would show only the favorable side of affairs.

On Dec. 6, Dr. Gruening, then with the *Portland Evening News*, warned

Special Haiti Commission, left to right, William Allen White; Elie Vezina; W. Cameron Forbes, chairman; Henry P. Fletcher and James Kearney.

that the fate of commissions usually is to be taken in by those they are investigating, and he declared that the first step toward a satisfactory solution must be the removal of the military High Commissioner, Gen. Russell, and that a civilian should appointed. The American Civil Liberties Union stated that the facts were already known, and that it was unnecessary to send a commission to investigate. It made a number of recommendations to alleviate the intolerable conditions alleged to have been imposed on Haitian citizens by the American Occupation.

The Greater New York Federation of Churches petitioned President Hoover expressing concern lest the United States violate the pledges made in the Kellogg Pact in its dealings with Haiti. The Colored Unity League of Jersey City accused President Hoover of complicity in the muzzling of free speech, unlawful imprisonment and probable murder then being inflicted on patriots in Haiti. Teachers should be sent there, not Marines, it said. Professors and students of New York's Union Theological Seminary sent a letter protesting the Occupation of Haiti as a violation of American ideals of liberty and justice, and charging that "American prices" were responsible for killing Haitians and preventing the growth of democratic government there.

On Dec. 8, 1929, President Hoover requested Congress to authorize and fund a presidential commission to study and review American policies in Haiti. Congress responded on Dec. 18 by passing a joint resolution (H.J. Res. 170) providing for such a commission and appropriating $50,000 for its expenses. No policy guidance was attached to the resolution.

President Hoover's request left the door open for congressional representation on the proposed commission, but the offer was rather studiously ignored. The feeling in Congress seemed to be that Haiti was a problem created by the executive branch, and therefore the executive branch should resolve it. There was remarkably little debate over the proposed commission in either the Senate or the House of Representatives.

Announcement that a commission would be formed brought on another flurry of White House mail concerning who should serve on the panel. Great pressure was applied to name a black commissioner. The NAACP sent a telegram saying that 12 million American citizens of Negro descent felt that they should be represented on any commission appointed. Oswald Garrison Villard of *The Nation* demanded representation not only by American blacks but also by Haitians. Many other such recommendations and demands were made.

In response to a State Department request, Gen. Russell forwarded President Borno's view on the matter. The Haitian president stated that he was not averse to a Negro serving on the commission but he had not changed his view that Haitians in general looked down upon American blacks. He further said he did not want to see an American Negro sent to Haiti in a treaty-related or U.S. government position.[1]

Appointing a black person to the commission was given serious consideration, but it appears the following confidential memorandum from Dr. L.S. Rowe, director general of the Pan American Union, was decisive in persuading Hoover not to name one:

1 Russell to Secretary of State, Dec. 25, 1929.

"I beg to recommend that no Negro be included in this Commission. The practice of appointing a Negro as U.S. Minister to Haiti has never been well received in that country and now that this practice has been abandoned, I hope that it will be possible to avoid a Negro on the Commission. Rightly or wrongly, educated Haitians regard themselves as far superior to the American Negro."[2]

On Feb. 7, 1930, President Hoover announced that the Haitian Commission would be comprised of W. Cameron Forbes as chairman, Henry P. Fletcher, Elie Vezina, James Kerney and William Allen White. He further announced that, "In cooperation with Mr. Forbes, I have requested Dr. R.E. Moton, president of Tuskegee Institute, on behalf of the Institute and such other educational affiliations as he may suggest, to undertake an exhaustive investigation into the educational system of Haiti, with a view to recommendations for the future."[3] Dr. Moton was one of America's most distinguished African-American educators.

Forbes, a wealthy Boston lawyer, had served in the U.S. administration of the Philippines from 1904 to 1913, the last four years of that period as governor general. In 1921 he had been co-chairman on the Wood-Forbes Commission to the Philippines. When Sen. McCormick, in 1924, recommended that a commission be sent to Haiti, Forbes was at the top of his list of recommended appointees. Forbes was a longtime friend of President Hoover.

Fletcher had a long and distinguished career in government service. In 1898 he was a private in Col. Theodore Roosevelt's Rough Riders, and later saw service in the Philippines. His diplomatic career included assignments in Cuba, China, Mexico, Chile, Belgium, Italy and Luxembourg. He was chairman of the U.S. delegation to the Fifth Pan-American Conference in Santiago, Chile in 1923, and was a delegate to the sixth in Havana in 1928. For a brief period during the Harding administration, he served as Secretary of State.

Vezina was a wealthy newspaperman. He was born in Canada, active in Roman Catholic organizations, and described himself as a student of Haitian affairs.

2 Dr. L.S. Rowe to Acting Secretary of State, Memorandum, Jan. 21, 1930.

3 Presidential Press Statement, Feb. 7, 1930.

Kerney was a respected newspaper editor and author who had written among other things a book titled "The Political Education of Woodrow Wilson."

White was best known of all the appointees through his prolific writing. He was the editor and publisher of the *Emporia Gazette* and a biographer of Woodrow Wilson. He was renowned as a champion of the "common man."

The inclusion of three newspapermen on the five-member commission appeared to be no accident. In the winter of 1929-1930, Haiti was a topic of frequent concern in the American press. Most editorial commentary was negative concerning U.S. policy in Haiti, and it may well have been the administration's hope to lower the journalistic voices raised against it by the character of the commissioners.

On Feb. 13, 1930, the Union Patriotique wrote to Sen. Borah requesting that he use his influence to have the U.S. government pay the expenses of legal counsel to represent the opposition in hearings before the commission. Borah forwarded the request to Forbes, who replied, "The Commission is not taking any counsel along and proposes upon arrival in Haiti to give all elements an opportunity to offer a fair presentation of their views. We understand that there are many very competent Haitian lawyers in Haiti."[4]

Advice and recommendations poured in to the commission during the two-week period between naming the members and their departure for Haiti, including a long memorandum from Emily Greene Balch. She warned the commission not to be led astray by those the Occupation pushed forward, but rather to seek out those of whom the American and Borno officials were most hostile and contemptuous.

Armed with this advice and a box of books and official reports concerning the Occupation, the commissioners sailed from Miami on Feb. 25, 1930, on board USS *Rochester*, which during the Spanish-American War was the battleship *New York*. It had served as Adm. Sampson's flagship at the battle of Santiago, where the United States first established dominance of the Caribbean.

4 Forbes to Union Patriotique, Feb. 22, 1930.

20

THE OCCUPATION ON TRIAL

What has the Department of Horticulture done for the development of our important native fruits like citrus, mangoes, avocadoes, coconuts and vegetables? Why the present head of this department, Mr. Leonard, spends 95 percent of his funds in landscaping and pays no attention on industrial fruits and vegetables? How many mules have been born in five years? What happened to the flock of White Leghorn and Rhode Island chickens brought in from the States? Who can patronize the dairy at the price milk is sold? How many among the American personnel can talk French enough to make themselves understood?

Anonymous complaints registered with the U.S. Haiti Commission.

The weather on the trip south through the Bahamas Channel was good. The captain entertained the commissioners at dinner and at motion pictures shown on deck. They seemed to enjoy the relaxing voyage, but nevertheless spent a great deal of time reviewing the material they had brought with them, and preparing the press release they intended to make on arrival in Port-au-Prince. With so many strong-willed and articulate individuals on the commission, this last was no easy task. The President's charge had been that "the primary question which is to be investigated is when and how we are to withdraw from Haiti. The second question is what we shall do in the meantime."[1]

The press release they eventually agreed upon was cabled, apparently more as a courtesy than anything else, to the State Department several days prior to their scheduled arrival in Haiti. It took the President's charge as a point of

1 Statement by President Herbert Hoover, Feb. 4, 1930.

departure, and stated the purpose of their investigation was to determine the progressive steps necessary for placing all activities of government in the hands of people chosen for that purpose by Haitian authorities; the withdrawal of United States officers and employees from Haitian service; and the termination at the earliest possible moment of the American Military Occupation. The commission's draft concluded with a promise to hear all sides of the controversy, in public or private hearings.

A great deal of labor went into careful choosing of words, and the commissioners undoubtedly were angered when, the day prior to landing in Port-au-Prince, Acting Secretary of State J.P. Cotton (whom Forbes probably considered his junior) cabled that the department did not desire to have their suggested statement released and provided instead one of its own. In a huff, Forbes replied that no statement would be released to the press on arrival.

The *Rochester* anchored at Port-au-Prince at 1:30 in the afternoon of Feb. 28, 1930, firing a 21-gun salute with the Haitian ensign at the fore. Fort National returned the salute. In top hats and tail coats the commissioners left the ship to the firing of a second salute of 19-guns (recognizing their ambassadorial rank) and the playing of the U.S. National Anthem by the ship's band.

The pier where the commissioners landed had been roped off and was clear of everyone except the official welcoming party. Outside of the pier enclosure and lining the route taken by the official automobiles, thousands of people were massed. The commissioners were cheered and there were many cries of "*Vive le President Hoover!*" A parade formed in their wake and huge banners were displayed which read: "Down With the Council of State!" "Long Live Senators Borah, Blaine, King, Shipstead and ALL Friends of Haiti!" and "End the Occupation!"[2]

The Haitian press gave extensive coverage to the arrival of the commissioners and to all their subsequent activities in the country. Biographies and pictures of each commissioner were published in almost every paper. In *La Presse* on March 3, 1930, a large front-page picture showed the commissioners in a formal pose in front of the American Legation. Despite the dust and heat

2 *The Baltimore Sun*, March 1, 1930.

USS *Rochester.*

of the Haitian capital, all were wearing top hats, tails and high-collared shirts. All carried canes.

During the commission's presence in Haiti, attacks on the Borno government and the American Occupation continued unabated in the opposition press. Editorials quoted everyone from Abraham Lincoln to Sen. Borah. The demands were always the same: "*Désoccupation!*" "*President d'Haiti élu par le peuple!*" "*Election legislatives!*" "*A bas le Martiniquais!*" [A reference to Borno whose father was supposed to have been born there.] "*A bas la misère!*" "*Respectez notre souverainté!*"

The first hearings before the commission were held on March 1. Typical is the following statement made in opposition to the Occupation.

"My voice is not strong enough, my words are not sufficiently strong, to express to you all the atrocities of the American Occupation in Haiti for almost fifteen long consecutive years. I apply to you, who are charged with healing wounds, which are still bleeding, the indignation of the Haitian people is very great....

"The American Occupation ruined our finances, introduced black misery in all the families, put the Haitian trade in bankruptcy and closed the doors to Haitian industries. The true citizens are discarded from public affairs; public offices are given only to those capable of any capitulations.... You who are guided by the laws of the starry flag, in the name of humanity and of the civilized people, kindly remove us from the humiliating yoke of this odious Occupation which deprives us of what is most sacred and most dear in the life of a people, that is its liberty and its independence."[3]

On Sunday, March 2, the opposition staged a parade of Haitian women for the benefit of the commissioners, but only William Allen White was present to observe it, the others either being away to see a polo match or at tennis. White stood with uncovered head outside the commissioners' hotel, the Excelsior, while the women filed by. More banners were displayed: "Sacred Heart of Jesus, Save Haiti!" and "Women of Haiti, Pray for Justice and Liberty for Haiti."

One of the marchers, dressed all in black, stopped in front of White and said: "Please, Commissioner, deliver us." As she spoke, she bowed. White took her by the hand, bowed, and as she backed away, touched his fingers to his lips and wafted a kiss. This act was followed by wild cheers and cries of "Long Live the Commission!" and "Long Live President Hoover!"[4]

Illustrating that in reporting, as in so many other things, the point of view of the observer is everything, William Allen White's gesture was interpreted in the Haitian press (in English) as follows:

"Our Colours Kissed by a Member of the Commission! Lafayette, here we are! shouted the American troops landing in France. The forgetfulness of Savannah is nearly redeemed: yesterday afternoon before the Commissioners' Hotel, one of its members respectfully touched his lips to our Flag presented by an aged woman. Our Colours kissed by a Representative of President Hoover! The greatest hopes cheerfully gallop in wanton hearts: our blue and red on American lips! Throughout the Republic the Haitian People will be sincerely

3 Statement of Salamon R. André, March 1, 1930.

4 *The New York Times,* March 4, 1930.

moved in learning Mr. White's healthful gesture."[5]

The "forgetfulness of Savannah" refers to the Haitian legend that black troops transported by the French from Saint Domingue helped fight the British at Savannah during the American Revolution. A few blacks apparently were taken there on French ships, but it's doubtful any were landed or engaged in the fighting.

Hearings in Port-au-Prince were held in government buildings and the rooms were always packed when public testimony was given. Years ago, while researching a paper, I interviewed an elderly woman who, as a young stenographer, had worked for the Forbes's Commission in Haiti. She recalled, "The very doors and windows of the hearing rooms were blocked by a sea of intent black faces. The air inside the rooms was stifling, and at one time or another every member of the commission's staff was ill."

It is not supposed that the commissioners themselves were immune to the "traveler's disease" in Haiti. The nights were so hot and humid that Forbes regularly left his room in the Excelsior and slept on board the *Rochester,* which remained at the commission's disposal throughout its mission.

In spite of the heat and discomfort, hearings were held every day except Sunday while the commissioners were in Port-au-Prince.

First priority was given to testimony by the opposition. The full commission ordinarily heard charges against the Occupation, while only a single commissioner frequently heard the Occupation's replies to those charges. This apparent bias in the proceedings was strongly resented by Gen. Russell and other treaty officials. Complaints to the commission and to the State Department largely fell on deaf ears, however. In an early but undated interview the following exchange took place between Constantin Mayard, ex-president of the Chamber of Deputies, and Commissioner Fletcher:

Mr. Fletcher: "Have the representatives of the United States tried to change the language and the religion of the people?"

M. Mayard: "Yes. Even though the French language had been recognized as the official language of the country, in all the departments the English language

5 *Le Nouvelliste,* Port-au-Prince, March 3, 1930.

only is used." [This, unquestionably, was a bald-faced lie.]

Mr. Fletcher: "Are the documents not written both in French and in English?"

M. Mayard: "The general's report is in both languages, but it is easier to obtain an English copy than a French one.... In public schools, English is used instead of French."

Mr. Fletcher: "A lack of tact."

M. Mayard: "Exactly."

Mr. Fletcher: "They have no diplomatic tact because they are soldiers."

M. Mayard: ... "All the economic measures taken by the American Financial Adviser and by the High Commissioner on taxes are prepared in full and submitted to be accepted by the Council of State. The taxpayer is not consulted."

Mr. Fletcher: "Taxation without representation."

M. Mayard: ... "It is the military occupation that is causing the trouble. If God and all the Saints were President, nothing would be well with the military occupation."[6]

On March 3, 1930, the editor and publisher of *Le Nouvelliste,* Ernest G. Chauvet, appeared before the commission:

Gov. Forbes: "I would like to know if you enjoy freedom of the press."

M. Chauvet: "Absolutely not. Though we have been able to print something since the commission arrived."

Mr. Kerney: "So we have done something for Haiti already...."

M. Leger: [Haitian counsel for the opposition] "How long did you stay in jail without any trial?"

M. Chauvet: "Once 40 days, another 16 days, and another 7 or 6 days."

Mr. Kerney: "Don't you have the right by *habeas corpus* to get out?"

M. Chauvet: "Not under martial law. The bond for pressmen is set at $5,000, that is the lowest bond, but for any other criminal, murderer, and all, the highest sum is $300 — any criminal and any murderer can get out for that but we have to put up $5,000."

6 President's Commission for Study and Review of Conditions in Haiti. Undated record of hearing.

Mr. Kerney: "They must think that newspaper men are rich...."[7]

French Catholic priests testified during the first week of hearings in Port-au-Prince, and complained about the amount of public support given the church during the period of the American Occupation. They said the funds provided were inadequate to pay the salaries of priests, particularly those outside the principal cities. Other complaints concerned taxes the church was required to pay. Recently the government had begun to levy import duties on goods brought into the country by the church. This was a particular irritant, as were instructions given in Service Technique schools by predominately Protestant American teachers.

The commissioners heard from spokesmen from dozens of opposition groups, each claiming to represent "all the Haitian people." A frequent warning placed in the transcript was that if the forthcoming presidential election were put in the hands of the Council of State, American machine guns would have to sink Haiti in blood. The activities of the opposition were not confined to speech making before the commission, however. Police reports from the city of Port-au-Prince tell of frequent political meetings and rallies. A typical meeting was that of the Haitian Young Peoples Patriotique League on March 4 where vicious personal attacks were made on President Borno and Gen. Russell, interspersed with the reading of poetry. At the conclusion of this meeting, an African-American newsman and his wife stood on a table for the purpose of counting the crowd. They were roundly applauded.[8]

Le Nouvelliste paid tribute to sympathetic representatives of the American press in Haiti to cover the hearings. One such was Helena Weed Hill of *The Nation*. In addition to attacking Marines for public drunkenness and other immoral behavior, she went after civilians in the treaty service as well.

"The country is teeming with merry young striplings, 'technical experts' who are receiving double and treble the amount their qualifications ever would bring them at home, while the real work is being done by underpaid Haitians. The native resentment at seeing the important positions filled by such

7 President's Commission for Study and Review of Conditions in Haiti.

8 Chief of Police, Port-au-Prince, to Commandant, Garde d'Haiti, March 5, 1930.

whipper-snappers with Haitians tolerated only in the lowest grades is easy to comprehend."[9]

From the record it appears that no treaty official appeared before the commission in the first week of its hearings in Port-au-Prince. Except for whatever mitigating influence informal conversations with American officials might have had, the picture painted for the commissioners was somber, threatening and decidedly anti-Occupation. Letters poured in from all sorts of strange organizations — from "Haitians Resident in Cuba," "The Student National Association of the Dominican Republic," "La Ligue de la Jeunesse Patriote," and many others. All carried the same message: After 14 years of occupation, the United States had failed to prepare Haiti for self-government.

9 Quoted in the *The Baltimore Sun,* editorial, March 16, 1930.

21

A POLITICAL SETTLEMENT

We informed the opposition leaders that we had been assured that M. Borno was not seeking re-election and was willing to accept a neutral non-partisan as his successor. We said we were willing to help find such a man. They all doubted that he could be found. M. Hudicourt asked with an air of finality: 'Where are we to find this white blackbird?' But they agreed to think it over and return the next day with their answer. They returned the following day and said they were willing to try to reach a compromise and asked us what we had to propose. We then suggested that, as they were the leaders of the combined opposition and claimed to speak and act for their followers, they should prepare a list of five neutral men, any one of whom they would accept as Provisional President. M. Borno would be asked to prepare a similar list. If one man's name appeared on both lists, he would be the merle blanc.

— Henry P. Fletcher, *Quo Vadis Haiti?* July 1930.

On March 6, 1930, Gen. Russell sent a confidential memorandum to Chairman Forbes referencing the adamancy and belligerent attitude of the opposition politicians, and warned of the danger of violence and bloodshed. He outlined a compromise plan originally suggested by Marine Col. Frank E. Evans, the commandant of the Garde d'Haiti. The plan called for the election of a provisional president, a political neutral satisfactory to both the opposition and the government, who would be responsible for calling legislative elections. The provisional president would agree in advance to step down in favor of a new president regularly elected by the new legislature.

The commissioners seemed favorably inclined to the plan, so Gen. Russell took it up with President Borno, who agreed to give it a try. He said that for psychological reasons the commission should make the proposal to the opposition. He assured Gen. Russell that no difficulty would be encountered in Haiti's Council of State.

Adopting the plan as its own, the commission drafted a long message to Washington on the evening of March 7. A rough draft of the message was addressed to "The Secretary of State, for the Confidential Information of the President." Forbes expanded the address to say, "The President, Washington, D.C." Apparently, he did not want anyone else in the department to further interfere with the commission's work. The text, marked "Strictly Confidential," follows:

"Your Commission finds the situation here critical, the public greatly inflamed and representatives of large groups of people organized in various patriotic leagues have appeared in public sessions and without exception stated that they would not accept any election of president by the Council of State, an appointive body which has acted as a legislature here for about 12 years and which has elected and re-elected President Borno. The demand made by practically all who have appeared, and supported as advisable by the Archbishop speaking for the church, is for the restoration of representative government by a legislature elected by the people and their choice of a President, which they say is the only way of satisfying the popular demand, and that any other course will not be accepted by the people and will be opposed with acts of violence. Officers of the Marines, the High Commissioner himself and Gen. Evans representing the National Guard, which is the Haitian police force, all agree the situation is tense and likely to result in bloodshed; in the present state of the public mind some small incident might precipitate serious consequences. With the approval of Gen. Russell, your Commission has convened a group of five leaders representing the various patriotic leagues, and represented to us as likely to be able to control the situation, and at our suggestion they have issued a statement calculated to calm the public mind.

"Further, it was suggested to them, after considerable discussion, that a compromise candidate for the Presidency be found — some man wholly out of politics, neutral and satisfactory to both sides, and willing to accept the election as President until a duly elected legislative body can be convened.

"On one hand, the objection of representatives of Patriotic leagues that they and the public would not recognize the election by the Council of State, has been met by their proposal to convene delegates from the country at large who will endorse and select the candidate agreed upon. On the other hand, after this has been done, the plan is for President Borno to secure the election of the compromise candidate by the Council of State. The new President is to agree to call the election of two legislative chambers as soon as possible and then present his resignation, permitting the National Assembly to elect the new President. This program has the complete endorsement of Gen. Russell, who believes that President Borno will accept it. The five representatives we have talked with are confident they can find the necessary candidate satisfactory to both sides and acceptable to the general public. This plan contemplates that only the candidate designated by the plan outlined above and elected by the Council of State should be recognized by the United States as the President on the understanding that the plan will be carried out in its entirety.

"In endeavoring to effect a compromise which will satisfy the popular demand and at the same time comply with the provisions of the constitution with regard to the election of a President, we have made it clear to the five representatives with whom we have discussed the matter that we are acting entirely unofficially. The Commission understands that Gen. Russell approves and is wiring the Department to this effect.

"If this plan meets with your approval, the Commission would appreciate a reply at the earliest moment, because of the danger of premature publicity, which is imminent. Time is very short, because the leaders desire to submit and receive from the public assemblies throughout the Republic the endorsement and approval of the compromise candidate before the fourteenth of April.

"The Commission is leaving for a tour of the Republic on Sunday morning but arrangements have been made to receive your reply promptly. Please reply through USS *Rochester.* Forbes, Chairman."[1]

The next day, March 8, Hoover wired Forbes as follows: "I entirely concur

1 Forbes to President Herbert Hoover, March 7, 1930.

in the recommendations of your telegram of March 7."[2]

On Sunday morning March 9, the commissioners left Port-au-Prince by motorcar for a trip across the country under the impression that the compromise had been accepted. Arrangements were made to release details of the plan to the press on the afternoon of March 9, and this was done. Monday morning, when they arrived at Hinche, they were surprised to learn that President Borno had sent a telegram to all prefects in the country branding as false reports that a provisional government would be formed, and stating that no legislative elections would take place prior to the nearest date fixed by the Constitution — Jan. 10, 1932.

A series of frantic telegram and telephone calls passed between Chairman Forbes and Gen. Russell. After some indecision, the commissioners continued the trip to Cap Haitian. Under pressure from Gen. Russell, President Borno capitulated on the evening of March 10. It may thus be seen that not only did Occupation officials propose the compromise plan eventually put into effect, but they negotiated its acceptance by the Haitian government. This was carefully concealed from the press, which gave all credit to the commissioners. Gen. Russell and his treaty officials pretended to have been taken by surprise when the plan was announced.[3]

The commission's reception at Cap Haitian was marked by more parades and more slogans. An organization of Haitian women presented Forbes with a red silk banner inscribed "God Bless President Hoover." Several of the commissioners found time to visit Christophe's Citadel.

In Washington, however, despite President Hoover's prompt approval, receipt of the plan for a political settlement touched off a flurry of activity in a clearly annoyed State Department. Objections were raised to what was interpreted as a "secret promise" to support the compromise candidate in advance of his election, and in the plan's failure to take into consideration the constitutional requirements concerning Haitian elections — particularly that for the president to name a date in "an even numbered year" for holding legislative elections.

2 President Herbert Hoover to Forbes, March 8, 1930.

3 *The Baltimore Sun*, March 10, 1930.

These objections were cabled to the commission. A reply marked "Confidential for the President and the Acting Secretary of State" was quickly put together. The commission argued that it was not making secret promises but merely taking action to redress grievances of long standing. The prompt restoration of representative government was held to be essential if fresh violence was to be avoided.

"For our own protection, we wish to know, before going any further, whether or not our government is going to insist that the new President of Haiti must wait until next year to call legislative elections to be held in 1932. If the Department intends to insist on this interpretation, the Commission is unanimously of the opinion that the compromise plan will be jeopardized and measures which the President would be reluctant to order may become necessary.

"We must earnestly submit that the Haitian Constitution was written and adopted to secure representative government and that no technicality affecting merely the time set for elections should deprive the people of their fundamental constitutional rights. Further action awaits your reply...."[4]

The message was signed by all of the commissioners. In Washington, it was thoroughly discussed by President Hoover and Acting Secretary of State J.P. Cotton. Final approval was cabled to Port-au-Prince on March 14. The commissioners then sent a message expressing gratification and stating their intention to leave Haiti on March 16. Cotton asked that the commission delay its departure until definitive agreements were secured from all political elements in Haiti. Chairman Forbes replied that he considered the work of the Haiti Commission to be at an end, that farewell calls had been paid, and that a delay in their departure would only hurt the situation. He said that, if special documents or signatures were desired, Gen. Russell could secure them. The commissioners departed as planned.

In the meantime, Gen. Russell had secured the names of five provisional candidates from the opposition, and from this list President Borno selected Eugene Roy, whose election by Haiti's Council of State was scheduled for April 14, 1930.

4 Forbes to the President and the Secretary of State, March 14, 1930.

Editorial comment in the American press concerning the settlement varied. *The Washington Post*, for one, was not persuaded:

"... The politicians of Haiti are a bad lot, and will never be satisfied until American influence has been wiped out. Since the treaty remains in effect until 1936, the United States must continue to exercise authority in Haiti, whether the native legislature functions or not. It remains to be seen whether the restoration of the legislature will improve the situation or not."[5]

The New York Telegram took a more sanguine view:

"The Haitian settlement is one of the quickest and most brilliant examples of diplomacy in our history. In method and aim it represents a reversal of 15 years of Yankee imperialism there.

"Unwilling to take on his shoulders responsibility for undoing the diplomatic and administrative blunders of preceding administrations, President Herbert Hoover appointed a commission to investigate and advise him.

"That in itself was enough to show the breakdown between the U.S. State Department and the Marine Corps regime, which is subject to the U.S. President but upon whose advice he apparently could no longer depend.... To meet the emergency the President gave the commission power to clean up the mess. It did so with the help of Haitian independence leaders and despite the actions of Dictator Borno."[6]

For better or worse, a political compromise had been struck that in effect gave the Haitian opposition everything it had demanded. While the compromise itself was suggested by the Occupation, the decision to meet the demands of the opposition was taken by the commission, and after hearing in depth but one side of the Haitian controversy. The bulk of the testimony and written reports of American treaty officials and the Haitian government, replying to the charges levied by the opposition, were presented to the commission after its return from the overland trip to Cap Haitian. Shortly before the commission left Haiti, 18 spokesmen for the government party, the National Progressives, appeared before the commissioners. One member of the group, Col. Auguste

5 *The Washington Post*, editorial, March 16, 1930.

6 *The New York Telegram*, editorial, March 17, 1930.

Nemoures, later informed Gen. Russell that Forbes told the group that it was unnecessary to make any statement, since the commission "had already made up its mind."[7]

In all, the commission spent 15 days in Haiti. Hearings at which the opposition stated its grievances were held before the entire commission, and the opposition was permitted to appoint counsel to attack American officials, who were not allowed to cross-examine their accusers or publicly refute the charges. This was particularly unfortunate inasmuch as the attacks and what was interpreted as the commission's concurrence therein were given wide publicity in both the Haitian and the American press. The High Commissioner's well-constructed and lengthy statement answering the charges was not made public because the commissioners considered it "controversial."

7 Russell to Department of State, No. 1713, May 26, 1930.

22

THE CASE FOR THE DEFENSE

Never in the course of our history has there been so much liberty, so much humanity, so much general safety, so much guarantee of life and property than under the present regime. No one dares dispute this fact.

National Progressive Party Memorandum, Port-au-Prince, March 7, 1930.

The other side of the story in Haiti, what might be called the defense of the Occupation, does not appear in the report later issued by the President's Commission for Study and Review of Conditions in Haiti, nor was there any effort to make that defense public. As previously shown, the defense was not permitted to influence the basic findings of the commission and this was, perhaps, as it should have been, given President Hoover's charge to the commissioners.

The decision to withdraw was made in Washington prior to the naming of the commission, and in that respect the "study and review" was purely pro forma. There is no doubt, however, that in failing to give Occupation officials an equal opportunity to place themselves on public record, a serious injustice was done to them, and the public was misled both as to the record of the Occupation and the probable effects precipitate withdrawal would have on the Haitian people.

Not logic and not humanitarian instinct governed, only political expediency. This was recognized and acknowledged privately in Washington. When Gen. Russell's rejoinder to the commission's report arrived at the State Department it was sympathetically received. A memorandum from Under Secretary of State J.P. Cotton to the assistant secretary for Latin American Affairs concluded, "I think this report should go to the Secretary and then be locked up, as under the present state of affairs I do not know of anything we can do about it."[1]

A common theme in the testimony of pro-Occupation spokesmen concerned the peculiar social conditions existing in Haiti both prior to and after the intervention. The absence of a middle class, the existence of a small but voracious elite, the illiteracy and ignorance of the overwhelming mass of the Haitian people, were presented as virtually insurmountable problems that had no short-term solution and which vastly complicated the constructive work of the Occupation. Measures taken to build a modern and progressive society in Haiti were the very measures that drew the most fire from critics of the Occupation.

The treaty department that suffered the greatest criticism from the Haitian opposition, for reasons briefly touched on earlier, was the Service Technique. Its director, George W. Freeman, appeared before the commission on March 13, 1930. He spoke of the necessity of creating in Haiti a skilled middle class, and of the difficulties he had encountered in attempting to do just that. Fully 90 percent of the Haitian people derived their living from agricultural and industrial pursuits, he said. Thus logic demanded that Haiti's educational system be geared to satisfy the needs of the overwhelming majority. This was where the controversy raged with the elite, however, for the limited resources of Haiti did not permit the establishment of industrial and agricultural schools without some decline in the support given the existing schools system.

In 1930 the Service Technique had enrolled more than 11,000 students, almost 8,000 of whom were in 65 rural schools. The problems in supporting even such a modest school system in Haiti were enormous. There was an almost total lack of trained teachers. When the Service Technique was established, only two Haitians could be found who had ever studied in an agricultural school. Malnutrition and absenteeism in the student body were serious obstacles. Staff members almost without exception had to be trained from the ground up, for educated Haitians would have nothing to do with the new schools system. In spite of this, at the time of the commission's visit to Haiti, the Service Technique staff was composed of 436 Haitians, 24 Americans and 16 men of other

1 "JPC" to White, undated memorandum [June?] 1930.

nationalities, mostly French.[2]

Technical schools director Freeman was burned in effigy and slandered in the Haitian press. His work in Haiti, which he believed in deeply, was about to be demolished. Freeman died in Port-au-Prince shortly after the commission's departure.

Another treaty service that came in for more than its share of criticism was the Garde d'Haiti. The charge most often heard was racial discrimination. Opposition politicians, particularly when addressing themselves to organizations such as *The Nation* and the NAACP, took the position that racial discrimination was unheard of in Haiti before the arrival of the Americans. They claimed that Marines serving with the Garde were overwhelmingly recruited from the southern part of the United States and therefore racially "tainted." Other criticism centered on the Marines' alleged lack of education and knowledge of the local language.

No people drew the color line more ferociously than the Haitians themselves. In 1930, at the time of the commission's visit, the best Haitian clubs were organized strictly on the basis of color. Even in the Garde, light-skinned Haitians refused to compete in athletic events with dark-skinned ones. These feelings were deeply ingrained and reinforced by memories of many bloody purges of mulattoes by blacks and of blacks by mulattoes in the past. The treatment afforded Haiti's "Syrian" community is an example of racial discrimination at its worst. The constitutional bar to white ownership of property in Haiti, removed in the American-dictated Haitian Constitution of 1918 but reinstated after the Occupation ended, is another fine example. These things were no secret to anyone even remotely familiar with the Haitian situation, yet time and time again the myth that it was the Americans who introduced the scourge of racial discrimination into the Black Republic was accepted at face value.

The rough draft of the commission's report contains the following penciled query, later lined out: "The colored line is drawn between the Haitians themselves. Should this be mentioned, and if so, how dealt with?" Gen. Russell discussed at length with Chairman Forbes the hypocrisy associated with racial

2 Statement of Dr. George W. Freeman, March 13, 1930.

prejudice in Haiti.[3]

On March 13, Commandant of the Garde d'Haiti, Major Gen. Frank E. Evans (Colonel, USMC), tried to answer the criticisms directed against Marines serving in the Garde. He furnished the following data in a "Compendium of Information on the Garde d'Haiti":

Residence in Southern States, 24; residence in other states, 92. Education Grammar School, 16; education High School, 51; education College, 49. Average length of service in Haiti, 4.09 years; average length of service in Garde, 3.58 years. Average length of service in military organization, 13.15 years. Language spoken, French, 85; language spoken, Creole, 92; language spoken, both, 80; no knowledge of either, 13.

In his testimony Gen. Evans defended at length the practice of assigning enlisted Marines to the Garde as junior officers, and the record of the Garde in "Haitianizing" its officer corps as fast as qualified Haitians could be graduated from the *Ecole Militaire.* There were at that time 77 Haitian and 115 American officers in the Garde. He also produced evidence which showed that, although Adm. Caperton's decree establishing martial law in Haiti was technically still in effect, trials before provost courts had practically ceased in 1923, there being none in 1924, two in 1925, four in 1926 and none since then.

In response to a petty complaint, he admitted that the six horses the Garde kept in Port-au-Prince were used occasionally for polo, in addition to ceremonial duties such as parades, escort for the president's automobile and so forth.

Gen. Evans listed as the Garde's current functions and responsibilities: Building its own barracks and outposts, fire protection, running the military school, communications, arms control and licensing, rural police, coast guard, repelling bandits, communal advice, prisons (until the 1920s the insane were also cared for in the prisons), making its own uniforms, repairing its shoes, making soap, first aid, showing motion pictures in the interior parts of the country and requiring attendance of children at schools.

On March 11, Financial Adviser and General Receiver Sydney de la Rue presented the commission with a general report on the operation of his service.

3 Russell to Forbes, March 13, 1930.

He responded in detail to all the charges laid before the commission by the opposition. The philosophy implicit in his report is typical of that expressed by nearly all of the treaty officials in formal testimony before the commission or in their written reports.

"I desire to record that, in the opinion of this office, the present political unrest in Haiti has been very largely caused by one condition and one only. The elite politician does not know how to work, will not be taught to work and does not desire to work. The newspapers giving accounts of the students of Damien in the November-December strike amply demonstrated this mentality. It is degrading to a Haitian aristocrat to work....

"There is no unrest in Haiti today among the peasants except as has been artificially stimulated and inculcated by political agitation, by false promises, by wild stories, rumors, lies and definitely subversive propaganda. Men have come before the commission whose testimony I have read and stated they represented thousands of people. This is a falsehood, and was cunningly stated to deliberately mislead the commission. No meetings had ever taken place, no roster of names had ever been prepared, and no actual society had ever been formed....

"The fact of the matter is that there are possibly less than one-third enough jobs in Haiti to support the elite. This means that once every so often the elite must change places or starve. There is not enough money in Haiti to support all of them without exploitation of the peasant or robbery of the public treasury. In this fact you have the explanation of the present condition, and you have the explanation of the future unrest that will occur in Haiti. The difficulty now is that the out-politician has been out a long time. He is likely to stay out a long time unless he can secure an immediate change. It is the last desperate throw of the dice to win back place, or to die; and some of them are old....[4]

Gen. Russell's report to the commission, not published as previously stated because it was "controversial," seconded the points made by the Financial Adviser-General Receiver in almost every instance. On the basis of the testimony and the documents given the commission, opposition charges of malfeasance

4 Report of the Financial Adviser-General Receiver to the Commission, March 11, 1930.

against treaty officials evaporate into unsubstantiated and petty gripes. The overall impression gained from the record is that the treaty services were in fact doing remarkable work with the barest of material support and in the face of open hostility from those they were supposed to be aiding and advising.

The most damning criticism of the Occupation, ironically, came from Gen. Russell himself. In a letter to Forbes on March 13, he stated his conviction that there were only two solutions to the Haitian problem — "Get in, or get out." By "get in" he meant the establishment of an American military government and an end to the hypocrisy of the advisory system. Only in that way did he see the possibility of creating in Haiti the political and social conditions that were the stated goals of the Treaty of 1915. He concluded that, "while tremendous advances have been made in the material rehabilitation of Haiti, and the happiness and prosperity of the mass of the Haitian people have decidedly increased, the Haitian people are today but little better fitted for self government than they were in 1915."[5]

5 Russell to Forbes, March 13, 1930.

23

THE COMMISSION'S FINDINGS

Their sophistication is charming but it is of the eighteenth century. Their pride has been justified by its works, but in the modern world it is doomed. The existence of this exotic society in the industrial civilization is a survival lovely but futile. Naturally, when the husky democracy of America meets all this caste pride, all this refinement of a departed French aristocracy in men and women of Negro blood, the race prejudices of the American whether he be a soldier or a civilian flares up. Instinctively he either sneers at what he sees or rejects what seems to be the vast amusing pretense of the mulatto rulers of Haiti. Bitternesses are inevitable. Understanding is tremendously handicapped. The wonder is that with our background we have done so well. Yet we may do better.

Deleted material from early draft of the Commission's Report, March 1930.

In Washington, the U.S. Commission on Haiti called as a body upon first the President, and then the State and Navy Departments. On March 28, 1930, President Hoover announced that he had accepted the commission's recommendations as the policy of the United States toward Haiti. The commission's report as published shortly thereafter reviewed the situation, and recorded its high praise of "General Russell's wholehearted and single-minded devotion to the interests of Haiti as he conceived them." It deplored the seeming ingratitude of Haitian politicians for the work that had been done, and found that the principal shortcomings of U.S. policy in the past had been "the failure of the Occupation to understand the social problems of Haiti, its brusque attempt to plant democracy there by drill and harrow, and its determination to set up a middle class — however wise and necessary it may seem to Americans."[1]

The formal recommendations and "sequent steps" to end the Occupation presented in the commission's report follow:

"The Commission recommends:

(1) That the detail of Navy and Marine officers for all Haitian services be made for a minimum of four years and that an effort be made to secure Americans who will agree to continue employment in these services, so that upon the expiration of the treaty a force of American doctors, engineers and police officers will be available for continued assistance to the Haitian government should it then desire it;

(2) That, if possible, some form of continuing appropriation for roads be urged for expenditure by the Haitian government, with a policy that will provide enough funds to keep all existing roads in suitable repair before any new construction is undertaken; also in regard to further construction, that only roads most urgently needed to develop regions now settled and under cultivation be undertaken until the present economic depression has passed;

(3) That the United States interpose no objections to a moderate reduction of the customs duties, internal revenue taxes, especially those imposed upon alcohol and tobacco, or to a reduction or elimination of the export tax on coffee, if the condition of the Treasury so warrants;

(4) That it be suggested to the Haitian Government that it employ one American adviser in each administrative department of the Government to perform such work as the respective Cabinet Minister may delegate to him, these officers to give expert advice and assistance to the Haitian Government, similar to that given by American officers in China, Siam and Nicaragua, for naval matters in Brazil, and for educational matters in Peru;

(5) That as an act of graciousness on the part of the United States, a moderate appropriation be made available during the continuance of the treaty to defray the cost of American civil officials in the Haitian Government service;

(6) That an appointment of a military attaché be made to the Legation when the time shall have arrived for a Minister to replace the High Commissioner, as

1 Report of the President's Commission for the Study and Review of Conditions in the Republic of Haiti.

the question of the preservation of order is of first importance and the Minister should have the advantage of his advice on military and police matters;

(7) That an adequate Legation Building be constructed immediately by the Government of the United States in the city of Port-au-Prince to provide a suitable residence for the American Minister and appropriate offices."

"Complying with your instructions to suggest sequent steps to be taken with respect to the Haitian situation your commission offers the following:

(1) That the President declare that the United States will approve a policy, the details of which the United States officials in Haiti are directed to assist in working out, providing for an increasingly rapid Haitianization of the services, with the object of having Haitians experienced in every department of the Government ready to take over full responsibility at the expiration of the existing treaty;

(2) That in retaining officers now in the Haitian service, or selecting new Americans for employment therein, the utmost care be taken that only those free from strong racial antipathies should be preferred;

(3) That the United States recognize the temporary Haitian President when elected, providing that the election is in accordance with the agreement reached by your commission with President Borno and the leaders representing the opposition;

(4) That the United States recognize the Haitian President elected by the new legislature, acting as a National Assembly, providing that neither force nor fraud have been used in the elections;

(5) That at the expiration of General Russell's tour of duty in Haiti, and in any event not before the inauguration of the permanent Haitian President, the office of the High Commissioner be abolished and a non-military Minister appointed to take over his duties as well as those of diplomatic representative;

(6) That whether or not a certain loss of efficiency is entailed, the new U.S. Minister to Haiti be charged with the duty of carrying out the early Haitianization of the services called for as the Declaration of the President of the United States above recommended;

(7) That as the commission found the immediate withdrawal of the Marines inadvisable, it recommends their gradual withdrawal in accordance with ar-

rangements to be made in future agreement between the two Governments;

(8) That the United States limit its intervention in Haitian affairs definitely to those activities for which provision is made for American assistance by treaty or by specific agreement between the two Governments;

(9) That the new Minister be charged with the duty of negotiating with the Haitian government further modifications of the existing treaty and agreements providing for less intervention in Haiti's domestic affairs and defining conditions under which the United States would lend its assistance in the restoration of order or maintenance of credit."[2]

The commission's report was received with dismay and bitterness by the High Commissioner. In his view, "The changes recommended by the Commission in its report are purely superficial and where they do not redundantly recommend steps already or long since in effect, constitute a weakening of our Haitian policy.... Such changes cannot but spell disaster to Haiti."[3]

In regard to sequent step (4) above, it was Gen. Russell's firm view that no election, before of after the intervention, had been held in Haiti by the Haitians without force and fraud, and that unless the United States government supervised the election, force and fraud would certainly be employed again. In the wake of the commission's departure, he saw political conditions worsening and at a near-revolutionary state, with only the continued presence of American forces preventing a return to pre-1915 norms.

The commission considered the work finished, however. It had outlined a plan for when and how the United States should get out of Haiti, and what should be done leading up to that time. It seems clear that from start to finish the commission was firmly in the hands of its chairman, W. Cameron Forbes. William Allen White, in a letter to Forbes written shortly after the commissioners left Haiti, seems to confirm this:

"... In this whole matter I was a boy at a husking, out of my medium. All my life I have been in politics, but as an agitator for causes. This commission was my first official effort. I could do nothing except furnish one vote out of five

2 Report of the President's Commission for the Study and Review of Conditions in the Rep. of Haiti.

3 Russell to Secretary of State, No. 1713, May 26, 1930.

in the commission, and perhaps my influence in two or three others. But the language of diplomacy, the forms of diplomacy all were not merely out of my experience but deeply different from anything I had ever known. It has been my job all my life to make myself as plain, as forceful, as direct and convincing as possible in order to effect public sentiment. The weapons for making and controlling public sentiment are fairly familiar to me and I have used them with some success during my 40 years in politics. But not one single weapon, not a rhetorical screwdriver in my whole kit and accouterment, was worth a tinker's resin dam on this job. But I hope I did back up Fletcher and you. I watched with 'wonder, awe, and praise' as the old song says, the deftness of your touch, the sureness of your tact, the beauty of your maneuvers. You can't teach old dog new tricks. I will never be of any use as a diplomat. But the look-in I got was worth the whole trip and I feel it is almost a shame to take my per diem in the face of my experience....

"I hear everywhere kind things about the work of the Forbes Commission. It is another jewel in your crown. I am proud to have been with you in this enterprise even if I could not contribute as much as I might have done in another kind of undertaking."[4]

Another letter caught up with the chairman in the last few days before he departed Washington for his home. It was dated March 21, 1930, and was postmarked from a small town in western Pennsylvania where an old man, retired from the missionary service, was living out his final years. The letter began:

"Dear Governor General:

"How I wished you could have taken me with you to Haiti, I could have been the very kind of help you required, absolutely confided in by leading Natives; yet an admirer of the High Commissioner, and best able to appreciate the special, and valuable reform brought about....

"We must not leave Haiti to herself. Do call me down dear Governor? Am positively certain your plan will be THE REDEMPTION OF HAITI. They are CHILDREN OF NATURE."[5] The letter was signed by the Rev. L. Ton Evans.

4 William Allen White to W. Cameron Forbes, April 9, 1930.

5 L. Ton Evans to W. Cameron Forbes, March 21, 1930.

24

HAITIANIZATION AND WITHDRAWAL

My great interest in Haiti and the Haitian people after spending a quarter of my life in Haiti for the purpose of assisting it, makes me strongly regret the tearing down of the constructive work which has been carried on in Haiti by the United States Government during the past fifteen years, and which has materially added to the health, prosperity, and happiness of the Haitian people.

— Brigadier Gen. John H. Russell, USMC, May 26, 1930.

On March 21, 1930, opposition delegates meeting in Port-au-Prince nominated Eugene Roy as their candidate for Provisional President of the Republic, in accordance with the agreement previously made. The Haiti Council of State was to have elected him on April 14 but as that date approached it became known there was serious opposition to Roy in the council. The opposition charged that President Borno was seeking to upset the agreement reached with the Commission on Haiti by failing to exercise sufficient influence among the council members. The American State Department warned Borno, through Gen. Russell, that the United States would recognize Roy and no other as provisional president.

On the eve of the scheduled election, President Borno abruptly adjourned the Council of State until April 21. Postponing the election created a serious situation. In Port-au-Prince there was an outbreak of arson that continued even after Roy's election on April 21. Roy's victory was assured after men pledged to support him replaced 10 of the 21 members of the council.

In the interim between election and inauguration, on May 15 the disturbances in Haiti deepened. The homes of some American officials, including that of the Marine Brigade commander, were damaged or destroyed by fire.

The Parisian Theatre in Port-au-Prince was razed. Threats were made by mail and by telephone to the lives and property of Americans and officials in the Borno government. Disorders spread to other Haitian cities, and in Cap Haitian members of the Garde were attacked by mobs. In these circumstances, President Borno requested the full exercise of martial law, but Gen. Russell, aware of the reaction such a step would have in Washington, refused.

Intelligence reports linked the Haitian arsonists to Bolshevik groups in the United States, particularly the Crusader News Agency and the American Negro Labor Conference.[1] More respectable organizations such as the NAACP and the ACLU petitioned President Hoover to force the immediate implementation of the Forbes' proposals.

In June 1930 the Commission on Education in Haiti, headed by Dr. Moton, arrived in Port-au-Prince. During a visit of several weeks, the commission gathered material for its report to President Hoover. It recommended the unification of educational systems in Haiti under the administration of the Haitian Ministry of Public Instruction, curtailment of the activities and expenditures of the Service Technique, and American assistance, primarily financial, in the establishment of an adequate educational system in the republic. The commission also gave its views as to the probable cause of what was increasingly recognized as the failure of the American Occupation:

"Had there been less of a disposition to deal with the island as a conquered territory and more to help a sister state in distress, less of a desire to demonstrate efficiency and more to help others to the efficient direction of their own affairs, less of enforced control and more of helpful cooperation, the United States today might have greater reason to be proud of her intervention in the affairs of a struggling neighbor."[2]

The Haitian government under President Roy soon lost the temporary popularity it enjoyed after its coming to power. The press attacked Roy personally for nepotism, graft and reluctance to set a date for legislative elections. Under pressure from the High Commissioner, however, he issued a decree on

1 Russell to Department of State, No. 1682, April 19, 1930.

2 Report of the United States Commission on Education in Haiti.

July 11 calling for elections the following October.

On Sept. 19, 1930, Gen. Russell instructed by letter all American officials in Haiti concerning the attitude of the U.S. government in the forthcoming legislative elections. Noting that neither the Treaty of 1915 nor any other agreement or protocol allowed the U.S. to assist in any Haitian election, he directed that strict neutrality be observed. He also ordered the Garde d'Haiti not participate in any way unless called upon by election officials to maintain order in the vicinity of voting booths.[3]

Early on the morning of Oct. 14, candidates at Port-au-Prince assembled their followers at respective headquarters, plied them liberally with rum and sent them to the polls in charge of selected leaders. This routine was repeated at voting places throughout the country. The elections were conducted without noticeable disorder. U.S. Marines were confined to their barracks. The majority of senators and deputies elected were from the ranks of those who actively opposed the Occupation.

On Nov. 1, Gen. Russell tendered his resignation as High Commissioner, and on Nov. 12, the day his post was officially abolished, he left the country. On Nov. 16, Dr. Dana G. Munro, ex-chief of the Division of Latin American Affairs, whose appointment had been announced in July, assumed his duties as U.S. Minister to Haiti.

On Nov. 19, the Haitian National Assembly met and elected a new president of the republic. He was Sténio Vincent who, ironically, had been president of the last national assembly when dissolved by Gen. Butler during the conflict over constitutional reform in 1917. Vincent was one of the most outspoken critics of the Occupation, editor of the foremost opposition newspaper, a founder of the Union Patriotique, and one of the authors and signers of the "Memoire" published by that organization in 1921.

The popularity of the Vincent government quickly waned, as had that of President Roy. Those members of the politically active elite who remained without an office in the new administration led a chorus of demands for the immediate dismantling of the Occupation. Instant Haitianization of the treaty

3 Proclamation of the High Commissioner, September 19, 1930.

services was the new "cause" for opponents of the Occupation.

Early in December 1930, President Vincent submitted a plan to the American minister calling for the complete Haitianization of Public Works in one year, and Public Health, the Service Technique and the Financial Adviser's Office in two years. No mention was made of the Garde.

The plan was considered "radical" by American minister Munro and the U.S. State Department. Against Munro's advice, an attempt was made to negotiate a new convention that would extend American financial control in Haiti, in the interest of foreign bondholders, while returning the other treaty services to the Haitian government. This received a lukewarm reception from President Vincent, but it met with cold hostility from almost everyone else.

Munro, who learned very quickly the difference between "being on the ground" in Haiti and being at a desk in Washington, reported that there was unquestionably a very strong desire that all Americans should leave at once regardless of the effect upon public order or the efficiency of government. It was clear that the process of withdrawing from the position hitherto occupied in Haiti would be an exceedingly difficult and disagreeable one. He wrote that he awaited a meeting of the Haitian Congress in April 1931 "with a feeling of dismay, which is only mitigated by the very evident terror with which the members of the government anticipate this same event."[4]

In the spring of 1931, Sen. King and Dr. Gruening, traveling with the Committee on Cultural Relations, visited Haiti. Enthusiastic crowds met these two patron saints of the anti-Occupation movement. In their speeches, however, they urged the people to exercise some restraint during the Haitianization process. This cost both of them popularity with the more radical elite. Attacks were made on them in the Port-au-Prince press because of their unexpectedly moderate stand.

When the new Haiti congress convened it proved to be as unmanageable and violently anti-Occupation as had been feared. One of its first acts was to vote its members an expense allowance of $100 per month retroactive to November of the previous year. The American minister, backed by Secretary

4 Munro to Secretary of State, No. 73, February 26, 1931.

of State Stimson, directed the Financial Adviser to make no payments under this new law and, as might well be imagined, this intensified the bitter feelings toward the Americans. Ominously, the congress began attempts to control appointments and promotions within the Garde. This was but the first step toward politicizing it and threatened a return of a military oligarchy.

A bad and deteriorating relationship with the Haitian government made it clear that the United States could not expect any cooperation in maintaining treaty services as efficient working organizations under Haitian control. U.S. policy therefore changed in the summer of 1931 to that which called for the turnover of American-directed services as soon as possible regardless of the consequences. A disillusioned Munro wrote, "I cannot see that it is any part of our duty toward Haiti to assist the Government in power by lending ourselves as a target for abuse."[5]

Pressure in the United States also encouraged acceleration of the Haitianization process. The ACLU sent a circular letter urging total American withdrawal; the NAACP demanded the appointment of a new commission, "at least one member of which shall be an American Negro," to devise steps for withdrawal of all U.S. control prior to the expiration of the treaty.[6] The so-called "Unemployed Councils of US" and the "Anti-Imperialist League" sent petitions. President Vincent appealed directly to President Hoover for more vigorous execution of the Forbes proposals.

In this atmosphere a Haitianization accord was signed as an executive agreement on Aug. 5, 1931. Martial law was formally repealed on that same date. Part of the accord provided that Haitian legislation no longer required approval of the American minister, and payments no longer had to be approved by the Financial Adviser.

In Washington the accord was hailed as a step made possible by remarkable progress achieved by Haitians in the management of their own affairs, a step that put into effect the recommendations of the Forbes Commission far ahead of the time thought to have been possible. It turned over to Haitians

5 Munro to Secretary of State, No. 159, June 26, 1931.

6 NAACP to President Hoover, July 9, 1931.

practically all of the treaty services, with the exception of the Garde and the Financial Adviser's Office. Financial control was reduced, however, to that considered essential for the protection of bondholders. Negotiations were even then under way that would lead to the purchase of the National Bank by the Haitian government and replacement of American responsibilities to bondholders by a civilian directorship.

The Haiti National Bank eventually was purchased for $1 million, but the bondholders refused to accept the second part of the plan. Consequently, the last vestige of the Occupation was destined to be a "financial representative" attached to the Haitian government who officially represented not the United States government but rather the bondholders. This arrangement continued throughout the 1930s.

In the fall of 1931 political activity in Haiti heightened, 1932 being another even-numbered year and legislative elections being called for again. In November there were press reports of political assassinations and vicious attacks on Americans as well as President Vincent. Late in November, irony of ironies, Vincent ordered the dissolution of the Union Patriotique, alleging it had departed from its original purpose to become a "political association."[7]

The elections were held on Jan. 10, 1932. In typical Haitian fashion, government candidates were elected in almost every district. A referendum attached to this election restored to the Haitian Congress, meeting in joint session, power to amend the Constitution. Shortly after the new congress convened in April, it exercised that power by extending its own members terms in office by two years.

The 1932 election of U.S. President Franklin D. Roosevelt was followed by a restatement of U.S. policy toward Haiti. It was now "to get the Marines out of Haiti as quickly as possible" because it was thought such a gesture "would make a good impression on South America."[8] Withdrawal of Marines would also, perhaps, still the clamor of domestic voices which, interestingly enough, were not lowered by the change of administrations in Washington.

7 Munro to Secretary of State, Cable, November 23, 1931.

8 Memo of telephone conversation between Francis White and Norman Armour, April 5, 1933.

Haitian President Sténio Vincent, left, meets with U.S. President Roosevelt, center.

On Aug. 7, 1933, the new American Minister in Port-au-Prince, Norman Armour, signed an executive agreement that provided for the complete Haitianization of the Garde d'Haiti, withdrawal of all Marines and inauguration of the "fiscal representative" by Oct. 1, 1934. Not long afterward, fresh Haitian demands were made for an immediate withdrawal of Marines and an end to all financial control, but not even Franklin Roosevelt was willing to accept this.

In April 1934, President Vincent traveled to Washington and there, on April 17, the presidents issued a joint statement concerning "the policy of the good neighbor" as it applied to Haiti. There were no significant modifications to the existing plan to withdraw, though Roosevelt did announce that he was requesting the Congress to authorize a gift of the Marine Brigade property to the people of Haiti. This property, valued at a modest $100,000, consisted mostly of weapons. The Congress later approved the gift, and President Roosevelt repaid President Vincent's visit on July 5, 1934.

Roosevelt arrived in Cap Haitian aboard the USS *Houston.* After brief ceremonies and discussions ashore, it was announced that, at President Vincent's request, command of the Garde would be turned over on Aug. 1 rather than Oct. 1 as previously scheduled. All American forces were to leave Haiti by Aug.15.

On Aug. 6 the last of the U.S. Marines withdrew from Cap Haitian following an impressive ceremony. The small detachment marched to the harbor accompanied by the Garde band, which played "Semper Fidelis," "The Stars and Stripes Forever," "The Halls of Montezuma," and finally, as the Marines boarded USS *Woodcock*, "Auld Lang Syne." It was reported there were tears in the eyes of many old Haitians as the Marines paraded by, and that not a few either took off their hats or saluted the American flag as it passed.

At 5:30 in the afternoon on Aug. 14, 1934, the American flag was lowered for the last time at Brigade Headquarters in Port-au-Prince. A company of the Haitian Garde faced a company of U.S. Marines on either side of the flagpole. As the American flag was slowly lowered, the Garde band played the U.S. National Anthem. A 21-gun national salute was fired at Fort National. The Haitian flag was then raised as the band played the Haitian national anthem. This was followed by a 21-gun salute to Haiti. The Marines then marched in formation from the compound and left at 9 a.m. the next morning on board USS *Argonne* and USS *Bridge*.

After more than 19 years, the American Occupation of Haiti had ended.

President Roosevelt traveled to Haiti aboard the USS *Houston*.

EPILOGUE

After the impressive ceremony of yesterday afternoon which took place in a perfect atmosphere of friendship, I am happy at the moment when the last Marines are embarking to renew to you the assurance of my gratitude, that of the Government, and of the Haitian people for your generous and intelligent policy of the good neighbor which has effectively aided me in accomplishing national liberation and which ensures the continuation of the cordial relations now existing between our two countries.

President Sténio Vincent to President Franklin D. Roosevelt, Aug. 15, 1934.

In keeping with Haitian political tradition, Sténio Vincent extended his term of office by unconstitutional means until 1941. In 1937 he very nearly had a war on his hands when large masses of Haitian peasants, starving and desperate, crossed the Dominican border seeking work. An estimated 10,000 Haitians were slaughtered on orders by President Trujillo of the Dominican Republic. He described the action as being necessary in turning back the "invasion." Later, the two presidents met and shook hands over the incident.

Street riots in Port-au-Prince in 1941 forced President Vincent to step down. The Haitian National Assembly then elected Elie Lescot. He, too, tried to extend his term by extra-legal means, but in January 1946 the Garde forced him out. A new National Assembly was convened and it elected Dumarsais Estimé, the first black (non-mulatto) president of Haiti since the dismemberment of President Sam in 1915. Estimé presided over the adoption of a new Constitution that restored the pre-1917 ban on foreign ownership of real property in Haiti. Under his leadership the Garde, renamed *Armie d'Haiti* in 1949, became the recognized political arm of Haitian blacks.

Haitian President Sténio Vincent, right, with his successor Elie Lescott, center.

In 1949 Estimé sponsored a world's fair in Port-au-Prince that drew many visitors to the country but also left it with a staggering debt. Denied re-election in 1950 by the National Assembly, Estimé called in black mobs to clear the chamber. Terrible riots swept through the capital, and finally the Armie d'Haiti removed the president and restored order.

Not unexpectedly, the next president of the Republic was the commander of the Armie d'Haiti, Col. Paul Magloire. He insisted on an election by the people, and in a festive spirit encouraged by free food and liquor, the people responded by electing him with a reported plurality of some 150,000 out of a total vote cast of 153,000. The early years of his administration appeared to be as good as any since the end of the Occupation, but in 1956 a general strike, rumors of excessive graft and corruption, and increasingly gloomy economic forecasts, led to his overthrow by the army he once commanded. He escaped to exile in New York taking with him the carefully hoarded cash reserves of the Haitian treasury, allegedly almost $7 million.

In September 1957 a respected country physician, poet and philosopher was elected Haiti's president. His name was Francois Duvalier, the beloved "Papa Doc" of the Haitian peasantry. In 1964 a new National Assembly, handpicked by Duvalier, elected him to a second term, even though his first was not due to expire until two years later. By referendum he was then declared President for Life. When he died in 1971, his son and heir, Jean-Claude "Baby Doc" Duvalier, 19, succeeded him. Baby Doc's appointment as President for Life was formalized by an even more blatantly rigged referendum. The published vote was 2,391,916 to 1, with two abstentions.

Thirty-seven years after the American Occupation ended, Haiti was the least democratic, least developed and poorest nation in the Western Hemisphere — and quite possibly in the entire world.

THOMPSON

26

PERSONAL NOTES AND OBSERVATIONS

It is a pleasure to stand upon the shore and to see ships tossed upon the sea; a pleasure to stand in the window of a castle to see a battle and the adventures thereof below: but no pleasure is comparable to standing upon the vantage ground of truth and to see the errors and wanderings and mists and tempests in the vale below.

— Lucretius (c. 99 - c. 55 B.C).

In 1965, while serving as commanding officer of a small Navy ship, I entered the harbor at Cap Haitian on the first of my several trips to Haiti. There is an excitement attached to entering a strange port, and after a night of cruising along a rocky and mountainous coastline the excitement was heightened a notch or two. During much of the night a bright Caribbean moon illuminated the island, and the raw smell of the land hung heavy in the air. Having read as a boy a few fanciful books about Haiti, my ears almost strained for the sound of voodoo drums.

The sea detail for Cap Haitian is a short one and the morning was perfect, with unlimited visibility, flat calm seas and only the slightest onshore breeze. Despite this, the approaches to the harbor were almost devoid of the throngs of fishermen and coastal sailors who clutter the entrance to most other ports in the world. As my ship slipped past Fort Picolet, the only navigational aid I saw marking the narrow ship channel was the broken mast of an old steamer jutting ominously from green water on our port hand. The buoys that once lined the way to the only usable deepwater berths had either sunk or been carried away over years of neglect.

Once moored, we were boarded by the chief of police and the only "official" American in the city — the director of a small USIA library in the center of town. The American consul recently had been ordered out of the country by President "Papa Doc" Duvalier.

After the usual exchange of pleasantries — social invitations, warnings about drinking the water and venereal disease, the scheduling of tours for the crew, etc. — I had a few free hours before calling on local Haitian officials. I decided to take advantage of that time and briefly explored the city. Everything worthwhile seemed to be in easy walking distance from the ship.

It then being siesta hour, the hordes of peddlers and beggars who greeted our arrival were asleep on the pier, stretched out nearby their drums, woodcarvings and brightly colored paintings. Only a few sinister-looking individuals at the pierhead appeared to be alert behind dark sunglasses. They were obviously members of Papa Doc's secret police — the *Ton Ton Macoute.*

Cap Haitian, the erstwhile "Paris of the Antilles," had the appearance of a dying slum. The city's narrow, dirt streets were potholed to the extent that wheeled traffic, where possible, appeared to be extremely hazardous. Sanitation was deplorable. Garbage and worse littered the streets and sidewalks. The one observable sign of civic pride was a plethora of tiny Haitian flags and banners inscribed *Doc à Vie!* fluttering pathetically from the fronts of mean little houses. President Hyppolite's Iron Market presented scenes that wrenched the heart. The market women, the *marchandes*, who likely had carried on swollen feet their pitiable produce to town over miles of stony trails, lay sleeping amidst stunted, half-spoiled vegetables and graying meats. All seemed oblivious to the flies, the stench and the noonday heat. A naked child nursed at his sleeping mother's breast. Older children, for the most part clad only in ragged shirts, threw stones at gaunt, starving dogs to keep them off the market floor.

This view of Cap Haitian may not be entirely a fair one. I later discovered that the green hills surrounding the city were dotted with small but charming homes. The newest, I was told, were built with American AID funds before the character of the Duvalier regime ended most U.S. assistance programs in Haiti. There was no attempt to conceal the fact, and no obvious embarrassment, that most if not all of that assistance in Cap Haitian had gone to benefit government officials. This was the natural course of events in Haiti.

The American Occupation was remote in time and all but forgotten. Most Haitians then living never even knew it had happened.

PRIMARY SOURCES

U.S. Government

Correspondence File of First Provisional Brigade of Marines in Haiti. Marine Corps Headquarters, History Section, Washington, D.C.

Department of State Numerical File, 1906-1910. U.S. National Archives.

Despatches from U.S. Ministers to Haiti, 1862-1906. U.S. National Archives, Micro Copy M-82.

Naval Records Collection, Subject File Haiti, 1911-1927. U.S. National Archives, Record Group 45.

Records of the Department of State Relating to Internal Affairs of Haiti, 1910-1929. U.S. National Archives, Micro Copy 610.

Records of the Department of State Relating to Political Relations Between the United States and Haiti, 1910-1929. U.S. National Archives, Micro Copy 611.

Records of the Naval General Board, Office of the Chief of Naval Operations, Naval History Division, Washington, D.C.

Records of the President's Commission for the Study and Review of Conditions in Haiti (Forbes Commission). U.S. National Archives, Record Group 220.

Records of the Senate Select Committee on Haiti and Santo Domingo, Medill McCormick, Chairman. Washington, U.S. Government Printing Office, 1921, 1922.

U.S. Department of State Decimal File 838.00, 1930-1939. U.S. National Archives.

U.S. Navy Logs, U.S. National Archives, Record Group 24.

Other

Beach, Captain Edward L. Sr. Papers (in possession of his family).

Caperton, Admiral William B. Papers, Library of Congress, Manuscript Division.

Forbes, W. Cameron. Papers, Harvard University, Houghton Library.

SECONDARY SOURCES

Books

Beach, Edward L., "The Wreck of the Memphis," Holt, Rinehart and Winston, New York, 1960.

Cave, Hugh B., "Haiti, Highroad to Adventure," Henry Hold and Co. New York, 1952.

Davis, H.P., "Black Democracy: the Story of Haiti," Dial Press, New York, 1928

Logan, R.W., "The Diplomatic Relations of the U.S. with Haiti," University of North Carolina Press, Chapel Hill, 1941.

Mahan, Alfred Thayer, "The Influence of Sea Power Upon History, 1660-1783." Little Brown & Co. (Twelfth Edition) Boston, 1942.

McCrocklin, James H., "Garde d'Haiti,"U.S. Naval Institute Press, Annapolis, 1956.

Montague, Luswell Lee, "Haiti ahd the United States," 1714-1938. Russell and Russell, New York, 1956.

Rodman, Selden, "Haiti, the Black Republic," Devin-Adair Co., New York, 1954.

Vandercook, J.W., "Black Majesty, the Life of Christophe, King of Haiti," Harper & Bros., New York, 1928.

Articles

Baker, Commander C.S.,"Some Colorful Haitian History," U.S. Naval Institute Proceedings, May 1924.

Boyden, Captain H.D., "The Curse of Santo Domingo", U.S. Naval Institute Proceedings, January 1937.

Coffey, Commander R.B., "A Brief History of the Intervention in Haiti," U.S. Naval Institute Proceedings, June 1922.

Conard, Captain Charles, "A Year in Haiti's Customs and Fiscal Service," U.S. Naval Institute Proceedings, April 1923.

Douglas, Paul H., "The American Occupation of Haiti," Political Science Quarterly, June and September 1927.

Fletcher, Henry P., "Quo Vadis Haiti?," Foreing Affairs, July 1930.

Greuning, Ernest H., "Conquest of Haiti and Santo Domingo," *New York Times* Current History, 1922.

Kelsey, Dr. Carl, "The American Intervention in Haiti and the Dominican Republic," American Academy of Political Science, 1922.

Seligman, Herbert J., "The Conquest of Haiti," *The Nation,* July 10, 1920.

Reports

Balch, Emily Greene, ed., "Occupied Haiti, being the report of six disinterested Americans representing organizations exlusively American, who, having personally studied conditions in Haiti in 1926, favor the restoration of the negro republic," The Writers Publishing Company, New York, 1927.

Bausman, F. & others, "The Seizure of Haiti by the U.S.," The Foreign Policy Association, 1922.

IMAGE CREDITS

Unless indicated below, all images are courtesy of The Library of Congress

Cover illustration and maps: Gill Guerry

Page	Source
28	U.S. Army
65	NavSource Naval History
69	Wikipedia
77	Wikipedia
79	NavSource Naval History
93	Naval History and Heritage Command
97	Naval History and Heritage Command
111	Naval History and Heritage Command
125	History Division, USMC University
129	History Division, USMC University
131	History Division, USMC University
229	NavSource Naval History
233	Bill Thompson

INDEX

A

B

C

D

E

F

P

R

S

T

U – Z

CUBA
Windward Passage
Tortuga
Port-de-Paix
Mole Saint Nicolas
Grande Riviere
Gonaives
Golfe
de la
Gonâve
Saint Marc
Île de la Gonâve
Jérémie
Port-au-Prince
Léogane
Petitgoave
Jacmel
Les Cayes
Isle-a-Vache
Caribbean Sea